Similarities in Physics

Similarities
in Physics

John N Shive

Robert L Weber

A Wiley–Interscience Publication

JOHN WILEY & SONS

New York

British Library Cataloguing in Publication Data

Shive, John N.
    Similarities in Physics.
    1. Physics
    I. Title   II. Weber, Robert L.
    530 QC21.2

ISBN 0–471–89785–7

Published in the U.S.A. by
John Wiley & Sons, Inc., New York.

Filmset by Macmillan India Ltd
Printed in Great Britain by Pitman Press, Bath

Contents

Preface

Physics is experience, arranged in economical order.

Ernst Mach

The great desideratum for any science is its reduction to the smallest number of dominating principles.

J Clerk Maxwell

If you are a science or engineering student who has already studied the first principles of physics, please imagine that we have written this book especially for you. Even if you are as far along as first-year graduate studies, you may find here some insights which will help you to tie together what you have already studied.

The edifice of physics is not nearly as complicated as it often appears to a beginning student. The purpose of this book is to tie similar phenomena together, wherever they may appear, in mechanics, acoustics, optics, electricity, heat, or theoretical physics and to call your attention to the similarities among them. By developing the habit of recognizing similarities, you can reduce the amount of material you have to learn. By developing a more comprehensive view of nature you will likely experience intellectual satisfaction and achieve confidence in dealing with new ideas, whether encountered in textbook problems or in real life situations.

In a little book *Similarities in Wave Behavior*, one of us (JNS) pointed out similar features of waves as they propagate, reflect, superpose, resonate, interfere, etc. Analogies were displayed among the behaviors of waves on mechanical, acoustical, electrical, optical, and thermal wave systems. The reader was invited, for example, to ponder how the non-reflecting coatings on the lenses of his binoculars are similar to the quarter-wave impedance-matching transformers on a wave guide transmission line. *Similarities in Physics* extends the philosophy of the earlier book: physics may be simpler than you think; nature is marvelously regular in its behavior; analogies helpfully link what is being studied to what is already understood.

You may find a bit of amusement and challenge in the following anecdote. Some years ago a recruiter for a large general electric company told us of a device he sometimes used during interviews, and about what it revealed. He said he would place a pen and a pocket knife on the table and then ask the person being interviewed to compare these objects.

The reply might be 'Well, the pen is green and the knife is brown. One has a point for writing, the other a blade for cutting. One seems to be made largely of plastic, the other of metal. One is longer and heavier than the other . . . ' and so on.

Another candidate might reply, 'The objects you are showing me are both tools. They are intended for hand use. They are of such size as to fit into a coat pocket. The cost is probably about the same for the two . . . ' and so on.

The first person has obviously been struck by *differences*; the second person has recognized *similarities*. 'Now,' our friend continued 'although I have made no formal statistical study, it is my observation of persons employed by our company that those who observe *similarities* very often turn out to be alert and innovative scientists. People who concentrate on *differences* may succeed well in other areas, such as sales and administration.'

Does this suggest a moral? Sharpen your ability to understand and apply what you observe and read in science by being alert to *similarities*. That's the keynote of this book! (Of course don't neglect to notice differences. Someday you may want to be chairman of the board.)

Although we did not originally set about to write a day-to-day teaching text, this book may be so used by teachers who wish to present second-level physics from the analogy point of view. Our primary purpose has been to develop insight. Some presentation of new knowledge is undertaken, to be sure, but chiefly to broaden the insight. The book may therefore find its greatest usefulness as collateral reading material. It should be helpful, too, to students at all levels studying for comprehensive examinations. We hope also that it will find some appeal to people who may read it just for pleasure.

John N Shive

Robert L Weber

Acknowledgments

The quotations at the beginning of the chapters are from the following sources:

Chapter 2   **Albert A Bartlett** 1976 *The Physics Teacher* **14** (7) 394

Chapter 3   **Harry F Olson** 1943 *Dynamical Analogies* (Princeton, NJ: Van Nostrand)

Chapter 5   **William Shakespeare** *Measure for Measure* Act V Sc I

Chapter 7   **George Bernard Shaw** *Man and Superman* III

Chapter 8   **Norbert Wiener** 1956 *I am a Mathematician* (Cambridge, MA: MIT Press)

Chapter 9   **Sir Robert Stawell Ball** 1872 *Wonders of Acoustics, or the Phenomena of Sound* (New York: Scribner, Armstrong & Co)

Chapter 11   **J B S Haldane** 1945 *Possible Worlds and Other Papers* (London: Chatto & Windus), first published in 1927

Chapter 12   **Eric M Rogers** 1967 *American Journal of Physics* **37** 692

Chapter 13   **Sir James Hopwood Jeans** 1942 *Physics and Philosophy* (Cambridge, UK: Cambridge University Press)

Chapter 14   **Pierre Duhem** 1954 *The Aim and Structure of Physical Theory* translated by Philip P Wiener (Princeton, NJ: Princeton University Press) p ix

Chapter 15   **Lucretius** *De Rerum Natura*, Book II, lines 141–44

Chapter 17   **George Herbert** *Man*

1 Generalized Steady Flow

Flow gently, sweet Afton, among thy green braes
I'll write an equation that tells of thy ways.
Thy sluggishness owes, as I'll prove in a while,
To a gradient shallow for mile after mile.

J N Shive *after* Robert Burns

1.1 The Linear Relationships of Nature

One of the tasks of the scientist is to discover how nature behaves, through observation, performance of experiments, and application of reasoning. Many generations of scientists, going about their work through decades and centuries past, have given us good descriptions of nature's regularities. What we already know fills thousands of books and journals, and the more we know, the clearer we can perceive patterns of consistency and universality which characterize the behavior of nature.

One of the tools employed by scientists is classification: things which are similar in one or more of their properties are grouped together for common consideration. If we can come to understand the behavior of one member of a class, then often we find that we are on the way to understanding the other members without having to initiate a detailed study of each one. Through classification we are led towards an understanding of the laws of nature which govern how things behave.

One of the broadest classifications in nature is that in which something is proportional to something else. Within large ranges of currents and voltages, the current in an electric circuit is proportional to the EMF, with the qualification that the current must not be so large as to cause substantial heating while the measurements are being made. As long as the elastic limit is not exceeded, the stretch of a spring is proportional to the force causing the stretching. The temperature rise of a beaker of water is proportional to the time of heating at a constant rate of heat input, and so on. These and a large number of other phenomena are describable by equations which say that the effect E is directly proportional to the cause C. That is,

$$E = kC, \tag{1.1}$$

where k is a constant of proportionality.

Such an equation is said to describe a linear relationship between E and C. There are other cases where something depends on the square root of some variable, or the square, or the logarithm, or the reciprocal, or the cosine and so on. How well behaved and considerate nature usually is in presenting us with laws expressible by such simple relationships! The linear relationship between two variables is the simplest and probably the most common characterization of nature. We shall begin our study by developing a topic of physics in which linear relationships prevail. This is the topic of steady flow.

1.2 Steady Flow of Electric Charge

In your first physics course you encountered the concept of electric current as flow of positive electric charge: the current I is the quantity of charge dQ per interval of time dt flowing past some point, in a wire for example. The current depends upon the potential difference between the ends of the wire,

$$I = \frac{dQ}{dt} = -\frac{(V_2 - V_1)}{R},$$ (1.2)

where R is the resistance of the wire. The minus sign indicates that the flow is directed away from the point of higher positive potential. This relationship you learned as Ohm's law.

If we remember that the conductance G of the wire is the reciprocal of its resistance, we can rewrite equation (1.2) as

$$I = \frac{dQ}{dt} = -G(V_2 - V_1).$$ (1.3)

This equation says that the current resulting from maintaining a potential difference between the ends of a conductor is linearly proportional to that potential difference, the constant of proportionality being the conductance of the conductor.

1.3 Ohm's Law in Other Guises: Heat Flow

Think of a metal rod of length l heated at one end and kept cool at the other end (see figure 1.1). The rod is packed in heat-insulating material so no heat is lost through its lateral surface. Experiment shows that the quantity of heat dH flowing in time interval dt

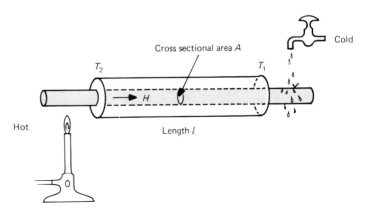

Figure 1.1 Heat flows from the hot end of the rod towards the cold end.

through a cross sectional area A of such a thermal conductor is proportional to the difference in temperature between the ends. That is,

$$\frac{dH}{dt} = -\frac{\kappa A}{l}(T_2 - T_1),\qquad(1.4)$$

where the factor κ is defined as the thermal conductivity of the material. The minus sign indicates that the heat flows in the direction opposite to that of increasing temperature.

Equation (1.4) is sometimes called Fourier's heat-flow equation. It is the conceptual analog of Ohm's law. However, it is heat which is flowing instead of electric charge, and the agency causing the flow is a temperature difference instead of a potential difference. Temperature, you see, has a kinship to potential; indeed, it is sometimes called thermal potential.

As a final step in developing the heat-flow equation we may combine the factors $\kappa A/l$ into a single quantity K, called the thermal conductance of the rod. We then have

$$\frac{dH}{dt} = -K(T_2 - T_1).\qquad(1.5)$$

1.4 Fluid Flow

A fluid flows through a tube because of a pressure difference between its two ends. In the case of a cylindrical tube of cross sectional area A, with the fluid flowing slowly enough to be non-turbulent, the volume dV of fluid passing a given point in the tube in the time interval dt is given by

$$\frac{dV}{dt} = -\frac{\pi r^4}{8\eta l}(p_2 - p_1),\qquad(1.6)$$

where r and l are the radius and length of the tube, η the viscosity of the fluid, and p_2 and p_1 the pressures at the ends of the tube. The minus sign indicates that the flow is from the higher-pressure end to the lower-pressure end. This relationship is called Poiseuille's equation after the French scientist who derived it in 1842 while he was studying the flow of blood through the capillary system of the body.

To rewrite this equation in such a form that it can be compared with equations (1.3) and (1.5), we may combine the factors

$\pi r^4/8\eta l$ into a single quantity F which we shall call the flow conductance of the tube. Equation (1.6) then becomes

$$\frac{dV}{dt} = -F(p_2 - p_1). \tag{1.7}$$

On comparing this equation with equations (1.3) and (1.5), you can see that the volume of liquid has a correspondence with the amount of electric charge and the quantity of heat, and that the pressure difference causing the flow of liquid is like the potential difference causing an electric current and the temperature difference causing a flow of heat.

1.5 Diffusion

In all gases, liquids, and solids the atoms, molecules, ions, and other particles are in motion. This motion, called thermal agitation, exists at all temperatures above absolute zero and becomes more violent with increasing temperature. It is responsible for the random Brownian motion of colloidal particles viewed with a microscope, for the thermal noise present in all electrical circuits, and for the diffusion of particles of one kind suspended in a medium of some other kind. Let's have a look at this last-mentioned process.

Suppose we consider a solution of sugar molecules in water, contained in a trough of cross sectional area A (see figure 1.2). Let us imagine further that the sugar molecules are not distributed with uniform concentration throughout the trough, but are more concentrated towards the left-hand end (where there are n_2 molecules per unit volume) than at the right-hand end (where there are n_1 molecules per unit volume). There is a uniform concentration gradient dn/dx from one end to the other. Because the particles of solvent and solute are in random thermal motion, there will be a net migration of sugar molecules away from the region where they are highly concentrated towards the region where they are less concentrated. After a time, then, the concent-

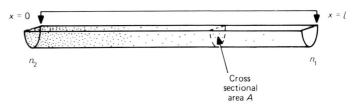

Figure 1.2 Solute molecules flow from left to right along the concentration gradient.

ration throughout the trough will become more uniform; the concentration gradient will become smaller and smaller, and ultimately will vanish as the entire trough of solution approaches complete homogeneity.

While this process is going on, there is a net current dn/dt of sugar molecules past a given cross section of the trough, given by

$$\frac{dn}{dt} = -\frac{AD(n_2 - n_1)}{l},$$

where l is the length of the trough and D is the *diffusion constant* of sugar in water. The minus sign indicates that the net flow is in the direction of decreasing concentration, against the gradient. If we call the quantity AD/l the diffusion conductance C of the trough, we can write

$$\frac{dn}{dt} = -C(n_2 - n_1). \tag{1.8}$$

1.6 Birds of a Feather

In §1.2–§1.5 we have looked at four transport situations, each one taken from a different subdiscipline of physics. If we gather the final equations from these four sections we have for comparison:

electricity	$\dfrac{dQ}{dt} = -G(V_2 - V_1),$
heat	$\dfrac{dH}{dt} = -K(T_2 - T_1),$
fluid	$\dfrac{dV}{dt} = -F(p_2 - p_1),$
diffusion	$\dfrac{dn}{dt} = -C(n_2 - n_1).$

$$\tag{1.9}$$

Can it be that the similarity of these equations corresponds to an underlying generality of the phenomenon of flowing? We assert that it most certainly does. Something is being transported, be it charge, heat, or whatever, and the magnitude of the flow is always directly proportional to the difference of potential, temperature, or whatever is causing the flow, as well as to the magnitude of the appropriate constant of proportionality.

If you were already familiar with two or more of the equations above, that familiarity was probably developed during your pursuit of several different courses taken in different terms. You were relearning the same basic concept, with similar equations, fitting the particular situations under study. Why, you may ask, was your developing acquaintance with flow phenomena made so repetitious? We can only answer that these different sub-disciplines of physics were developed at different times, by different investigators possibly unaware of what had been done in other areas; and, as so often happens, when a subject is developed in bits and pieces, that's the way it may be taught for generations to come. The flow equations were originally presented to you in piecemeal fashion because that's the way they were unfolded to your teachers.

The purpose of this book, if we may restate it now that we have made enough of a beginning for you to see what it is to be about, is to point out how the recognition of similarities can simplify your study of physics and give you a more comprehensive insight into the ways of nature.

One of the proper tasks of a scientist is to be alert to possible ways in which some new phenomenon he encounters may be similar to another phenomenon he already understands. If he can recognize such similarity, he may be spared much of the labor of studying the new phenomenon in detail. What he already knows of the previous phenomenon will guide his study, suggest to him what to look for, and possibly lead him to valid conclusions with much less effort than would be needed if he had to embark on a complete and thorough investigation of the new phenomenon, starting from scratch.

1.7 Applications

Modern technology abounds with examples demonstrating the similarities suggested by equations (1.9). Electrical network simulations of the Mississippi River System are used to predict the amplitudes and times of flood crests at locations downstream from a storm region. Thermal conductivities of materials are determined by a technique identical to that for determining electrical conductivities: heat is made to flow along a rod of the material and the resulting temperature gradient is measured. In the heating systems of many homes, hot water is circulated through pipes and radiators with flow maintained by a pump,

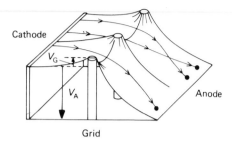

Figure 1.3 A rubber-membrane model is used for experimental determination of electron paths. V_A, anode potential; V_G, grid potential. (After *Introduction to Electron Microscopy* by C E Hall © 1953 McGraw-Hill, used with the permission of the McGraw-Hill Book Company.)

which is analogous to a DC electrical circuit where a current is maintained by an electron pump called a battery.

A mechanical analogy has been used to predict the paths of electrons in various types of vacuum tubes and in the electrostatic lenses for an electron microscope. The experimental method is suggested by figure 1.3, which represents a plane cathode, an anode, and a system of parallel grid wires in between. The height of the electrodes (pegs in the model) above a horizontal plane is in proportion to their negative potentials. When a rubber membrane (similar to the rubber dam used by a dentist) is stretched over the model it assumes a shape such that the height at any point between electrodes is in proportion to the electrostatic potential. The gravitational potential in the model is thus the analog of electrostatic potential. Electron paths are predicted by releasing small steel spheres at the cathode and allowing them to roll under gravity, as shown in figure 1.3. A camera with open shutter placed above the model records the paths of the spheres, as in figure 1.4.

Many of the control and information-processing functions performed by electronic devices can also be performed by analogous devices using fluid dynamic phenomena, without use of any moving mechanical parts. Figure 1.5 represents a fluidic device analogous to a three-element valve. It uses a stream of gas or liquid instead of electrons. The grid is replaced with precisely controled fluid jets perpendicular to the main flow. The control jets can function to modulate or deflect the fluid stream (fluid amplification) or to switch this stream 'on' and 'off'.

Almost any fluid can be used to operate fluidic devices. The devices can be made to operate in adverse environments of temperature, vibration, and severe radiation fields such as those

Figure 1.4 The electron paths in a triode are simulated with the rubber-membrane model of figure 1.3. A, anode; C, cathode; G, grid. The grid potential is increasingly negative from A to C. (After Kleynen 1937 *Philips Technical Review* **3** 338, courtesy of Philips Research Laboratories.)

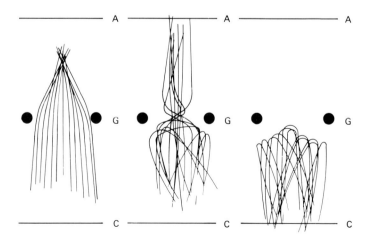

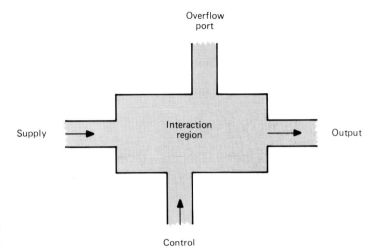

Figure 1.5 This fluidic device is the fluid analog of an electron triode valve.

produced by nuclear detonations, and they have therefore found specialized applications in spacecraft, jet engines, railroad loco-motives, and nuclear reactors.

In figure 1.6 the unit on the left is a four-input 'or/nor' gate; the unit on the right is a four-input flip-flop. The flip-flop consists of a solid block with fluid channels cut in one surface and covered with a transparent top to confine the fluid flow to the channels. The fluid stream introduced at the right may exit arbitrarily at

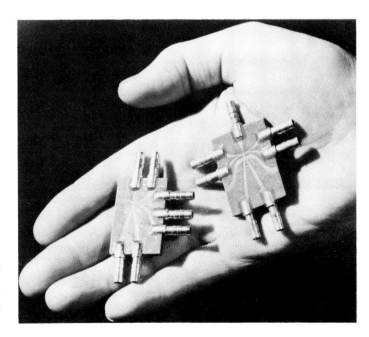

Figure 1.6 These are working models of fluidic control units: (*left*) OR/NOR gate; (*right*) flip-flop. (Courtesy of Corning Glass Works, New Materials Department.)

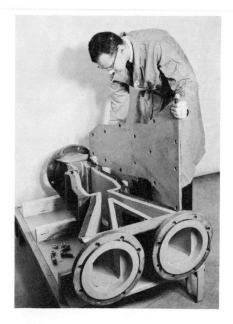

Figure 1.7 Fluidic devices can be impressively large. (Courtesy of Moore Products Company, Moore Industrial Controls.)

either left-hand port. A control signal at one of the top or bottom ports can divert the stream. The output remains at the last port used even when the controling signal is removed because of the wall-hugging nature of the flow. It can be said that the flip-flop has memory.

While compactness is characteristic of many fluidic devices, not all are tiny, as is evident in the custom-built fluidic valve shown in figure 1.7.

▶ **Exercises**

1.1   Discuss and give some other examples of linear relationships not brought out in this chapter. Can you think of any physical processes describable by equations which are *not* simple relationships like those of §1.1?

1.2   Describe the dispersal of a New Year's crowd from Times Square after midnight as a case of diffusion along a concentration gradient. How does such a dispersal differ from 'pure' thermal diffusion?

1.3   Work is done in an electrical system when charge is transported against an electric field. Describe how work is done in gravitational and hydrodynamic systems by analogous processes.

1.4   The force on a charge Q in an electrical potential gradient is $Q \, dV/dx$. By analogy, what is the force on a volume of liquid in a pressure gradient? Why is this analogy not applicable to describing the force on a quantity of heat in a temperature gradient?

1.5   Point out some similarities and differences among (a) flow of water through a horizontal pipe of circular cross section and variable diameter, (b) flow of air through a similar pipe, and (c) an electric current (DC) in a solid conductor of circular cross section and variable diameter. (Water

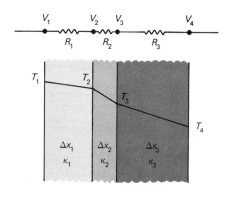

Figure 1.8

and electric charge are to be regarded as incompressible; air is compressible.)

1.6   If a certain physical property, say the resistance R of a lamp filament, depends upon some condition, say its absolute temperature T, even if this relationship is found experimentally to be complicated, we may be able to describe it mathematically if we use enough terms in a power series: $R = a + bT + cT^2 + dT^3 + \ldots$, where a, b, c, ... are constants determined from experimental data. To a first approximation (often quite satisfactory over a small range of T) we can disregard higher order terms, $T^2, T^3, \ldots$, leaving $R = a + bT$, a linear relation. Is this all that §1.1 is saying? Explain.

1.7   The current through a series of three resistors (figure 1.8) is given by

$$I = \frac{V_1 - V_4}{R_1 + R_2 + R_3}.$$

Develop the thermal analogy and show that the heat flow through a wall of area A composed of three plane layers of materials having thicknesses Δx_i and thermal conductivities κ_i (for $i = 1, \ldots, 3$) is given by

$$\frac{dQ}{dt} = \frac{T_1 - T_4}{(\Delta x_1/\kappa_1 A) + (\Delta x_2/\kappa_2 A) + (\Delta x_3/\kappa_3 A)}.$$

1.8   A cylindrical pipe made of ceramic material of high resistivity ρ and length l has inner and outer radii r_1 and r_2 (figure 1.9). Both cylindrical surfaces are covered by layers of ideal electrical conductor. When a potential difference V is applied between these layers a small current is produced through the pipe wall. Prove that the resistance R of this pipe between inner and outer surfaces is

$$R = \frac{V}{I} = \frac{\rho}{2\pi l} \ln\left(\frac{r_2}{r_1}\right).$$

1.9   Suppose that the pipe of Exercise 1.8 is made of material having thermal conductivity κ and that the inner and outer surfaces are maintained at temperatures T_1 and T_2, where $T_1 > T_2$. Show that the 'thermal resistance' of this pipe is

$$R = \frac{1}{2\pi\kappa l} \ln\left(\frac{r_2}{r_1}\right).$$

(If we replace the conventional thermal conductivity κ by its reciprocal, thermal resistivity ρ, this equation takes the same form as that for Exercise 1.8.)

1.10   The term 'one-dimensional' refers only to the number of coordinates needed to describe the distribution of temperature (or potential, etc) in a body, not the number of space dimensions occupied by the body.

Consider (a) a steam pipe covered with a thick coaxial layer of asbestos insulation and (b) a long fluorescent light, tube. Derive equations showing the role of radial distance r for (a) the heat flow per unit area within the insulation and (b) the illuminance (flux per unit area) from the lamp.

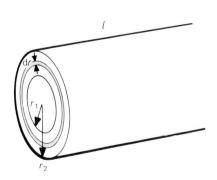

Figure 1.9

2

Exponential Growth

The greatest shortcoming of the human race is man's inability to understand the exponential function.

Albert A Bartlett

2.1 Introduction

We want next to consider a large class of phenomena in which a growth occurs over some time, not at a constant rate but at an ever-increasing rate so that the output of the process increases faster than it would at a constant rate. Nature provides many examples of such processes for which the rate of growth at any instant is proportional to the amount present at that instant.

The growth of a savings account when left in the bank at compound interest furnishes an example. So does the normal growth of a population in which the annual growth is proportional to the size of the population. Likewise the growth with time of an avalanche in which the mass of new material joining the avalanche each second is proportional to the mass already in the avalanche. The same kind of runaway build-up occurs even for the spread of gossip in a society if everybody who hears the gossip passes it on to two additional people every 24 hours. Growth processes exhibiting these characteristics are called exponential processes.

2.2 Compound Interest

As an example of exponential growth let us consider a savings account into which a depositor makes a single initial deposit of $\$_0$ dollars. He allows the annual interest to remain in the bank to swell the amount year after year. He wishes to know how large his nest egg will have become at the end of n years.

The annual interest on the account in any particular year is directly proportional to the size of the account in that year. That is, $d\$/dt \propto \$$, where $\$$ is the size of the account at any time. The constant of proportionality in this expression is the interest rate, r, defined as the *fractional* amount by which the account increases per year. We can thus rewrite the proportion as an equality, $d\$/dt = r\$$. When we separate the variables so $d\$/\$ = r\,dt$ and integrate, we obtain $\ln \$ = rt + c$, where c is the constant of integration. To evaluate this constant we recall that when the account was started at $t = 0$, the initial amount was $\$_0$. Hence $\ln \$_0 = 0 + c$, or $c = \ln \$_0$. Finally on recasting this equation into exponential form we have

$$\$ = \$_0\, e^{rt}. \qquad (2.1)$$

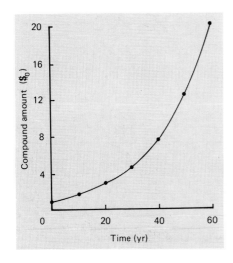

Figure 2.1 Money invested at 5% per year grows exponentially by geometric progression.

Figure 2.1 is a graph of equation (2.1) calculated for an interest rate of five per cent per year ($r = 0.05\ \mathrm{yr}^{-1}$). This kind of plot is called an *exponential* growth curve. One of its properties is that the growth continues without limit, increasing year after year by an accelerating amount as long as the initial specification of the conditions remains unchanged. Another property is that the amount increases by equal *ratios* in equal time intervals, no matter where on the curve you choose those intervals. That is, the increase is by geometrical progression.

▶ **Example 2.1** You may recall that when you studied compound interest in school, you learned the formula for an account started with $\$_0$ at an interest rate r for n years as: $\$ = \$_0 (1 + r)^n$. If so, you may be saying to yourself that $(1 + r)^n$ must give the same result as e^{nr}. Actually it doesn't, quite. You might take as a practical example the present value of an account started with $1000 and continued for four years at five per cent. By the binomial series expansion of $(1 + 0.05)^4$ you get $1215.51, whereas by a Taylor's series expansion of the exponential formula you get $1221.40. Why the difference?

The formula $(1 + r)^n$ assumes that the interest is calculated and added to the account *once* a year at the end of the year. On the other hand, the exponential formula assumes that the interest is compounded *continually* every day, minute, second.

One can make an adjustment to the $(1 + r)^n$ formula to make it correspond to the continual compounding of interest thus: $(1 + r/q)^{qn}$, where q is the number of times per year the interest is compounded. Then, as you permit q to become very large, this formula will give the same result as the exponential formula. In fact, it might be a good exercise for a degree candidate in mathematics to prove that $(1 + r/q)^{qn} = e^{rn}$ as $q \to \infty$. Why not try it?

A particular feature of an exponential growth is the doubling time, defined as the time interval required for the quantity to double. From the exponential curve of figure 2.1 one can see that the doubling time for a savings account at five per cent is about 14 yr. In another 14 yr it will double again, and so on. To become rich, then, you have only to invest a few dollars as a child and reap your millions when you grow old. Every dollar invested at five per cent when you were 10 years old will get you 32 dollars when you are 80. If you then bequeath this bonanza to your 10-year old great-grandson, and he continues the account until *he* is 80, he will have a $1024 windfall. Let's hope that when it comes due to him about the year 2110 it will at least buy him a loaf of bread.

▶ **Example 2.2** A very large sheet of paper 6×10^{-3} cm thick is folded in half, then folded in half again, etc. How thick will the sheet be when it has been folded 25 times?

Each time the paper is folded, the thickness d doubles. Therefore, after 25 folds,

$$d = 2^{25} \times 6 \times 10^{-3} \text{ cm}$$
$$\simeq 2.0 \times 10^5 \text{ cm} = 2.0 \text{ km!}$$

2.3 Population Growth

The compound interest problem offers an idealized example of pure and precise exponential growth: the exact conformity with an exponential solution was programmed into the situation by the man-made specification of constancy of the interest rate. In nature, cases of exponential growth are often approximate and prevail only over limited periods of time. As an example of such an approximate situation let us consider population growth.

Suppose the growth obeys the law that the amount of growth in any year is proportional to the size of the population in that year. That is, suppose $dN/dt \propto N$ or $dN/dt = gN$, where N is the number of individuals in the population and g is the overall growth rate, expressed as the ratio of the increase per year to the size of the population at the start of that year.

Solution of this differential equation follows that of the text in the previous section with the result

$$N = N_0 e^{gt}, \qquad (2.2)$$

where N_0 was the initial size of the population at the time $t = 0$, when we started counting time.

Biological populations generally exhibit truly exponential growth only over limited periods of time, and there are several ways in which they may deviate from it. In the first place, the exact proportionality between growth and population size is not always strictly found in practice. For example, in a human population, a war will deplete the ranks of young males, leaving the surviving population with a temporary imbalance of females of child-bearing age. In the second place, the overall growth rate g is strictly constant only over limited portions of the life of the population. This is hardly surprising when we recognize that the overall growth rate is the algebraic sum of the birth rate b, the death rate d, the immigration rate i, the emigration rate e, and the predation rate p. Thus,

$$g = b - d + i - e - p,$$

and changes in any of these terms affect the overall growth rate.

As an example, a population expanding in an environment providing only a limited food-supply rate will eventually over-run its support. Some individuals will then starve, thus increasing the death rate and slowing down and ultimately completely arresting the growth. Some animal populations, say of rats on the city garbage dump, will attract few predators when the population is small, but as it grows it will become increasingly interesting to hawks, dogs, and small boys with air guns and sling shots. The predation rate will therefore increase, further lowering the overall growth rate, sometimes to the point of zero growth. Biological populations exhibit truly exponential growth only for those limited intervals of time during which the overall growth rate remains stable.

In many cases populations may diminish in size. Consider a colony of bacteria growing in a culture medium providing only a fixed and limited amount of nutrient. As long as the medium provides nutrient in quantities which are more than sufficient for the size of the population, the growth will be purely exponential. However, when the available nutrient begins to run out, the population growth is arrested. Unless the nutrient is replenished, the death rate by starvation will thereupon dominate the overall growth rate, which will then become negative in sign, and the population will wither away. Figure 2.2 shows such a population rise and subsequent decline. The exponential growth portion of this profile represents only a small fraction of the total picture.

One way to tell whether a particular process is truly exponential or not is to plot the growth on semilogarithmic graph paper and see whether the result is a straight line. It doesn't matter whether one uses natural or decimal logarithms for this plotting; if the growth is exponential, the plot will be a straight line in either case. The growth rate g can be obtained from the slope of the line. If the semilogarithmic plot uses natural logarithms, the growth

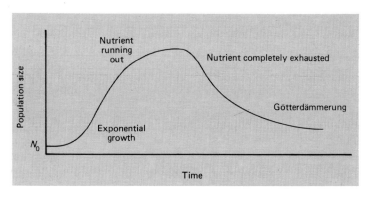

Figure 2.2 The growth of a colony of bacteria is checked by a limited amount of nutrient.

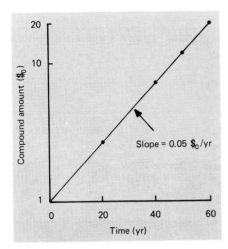

Figure 2.3 An exponential growth gives a straight line on a semilogarithmic plot (cf figure 2.1).

rate will be equal to the slope. If decimal logarithms are used, the growth rate will be 2.303 times the slope. Figure 2.3 shows the savings account data of figure 2.1 replotted on semilogarithmic coordinates.

As a practical case in point, figure 2.4 shows a semilogarithmic plot of the population growth of the United States from the time of the first census in 1790 to the present. You can see that this growth curve is not a straight line. However, it does have two clearly distinguishable straight-line segments, within each of which the growth is nearly exponential, with a decided change in the growth rate occurring around 1900. The doubling time of the early segment of this growth plot is about 24 yr; that for the more recent period is around 55 yr.

To explain this change in rate we may recall that in the first century of the Republic the immigration rate far exceeded the natural reproductive birth rate of the American people. In 1883, for example, over 750 000 immigrants entered the country when the overall growth was only 1 200 000 for that year. During the last 75 years, however, immigration has been greatly reduced, leaving the normal birth rate as the principal contributor to growth.

2.4 Growth of Science

Human enterprises, for example science, may grow exponentially. Dr Derek J de Solla Price has shown that if we measure the 'size' of science either by the number of scientists, or publications, or money spent, there has been an exponential growth (see figure 2.5). In countries where scientific development started later than in Europe, the rate of growth has been more rapid. It is

Figure 2.4 The growth of the population of the USA shows two exponential segments.

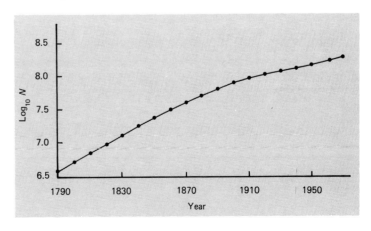

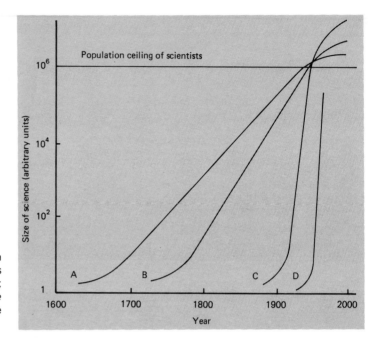

Figure 2.5 This schematic graph shows the rise of science in various regions of the world: A, Europe; B, USA; C, USSR; D, China. (After *Science Since Babylon* by D de S Price © 1975 Yale University Press.)

challenging to speculate on the effect of the predicted merging of these curves on international cooperation or tension and on the value judgments that will have to be made in science when we can no longer increase money and man power to investigate every problem.

2.5 Impact Ionization

In many processes the purely exponential growth of the type predicted by equation (2.2) and by the broken curve of figure 2.6 is modified by limiting factors to give a growth curve which approaches a saturation limit.

Consider a volume of gas in which two parallel-plate electrodes are placed. A potential difference is maintained between the plates by a variable voltage supply in an external circuit. Such a device, sketched in figure 2.7, is called an ionization chamber. If a source of x-rays is brought into the vicinity of the ion chamber, the x-rays will ionize molecules of gas in the space between the electrodes, and the electric field will cause the ions to migrate to the electrodes and give up their charges to the metal plates. The resulting current in the circuit is then a measure of the intensity of the x-rays.

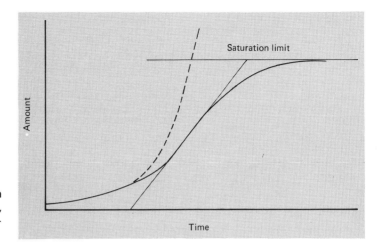

Figure 2.6 Exponential growth is often followed by saturation. Broken line, purely exponential growth; full line, exponential growth with saturation.

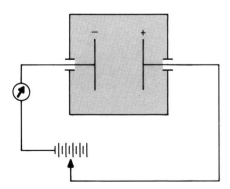

Figure 2.7 An ionization chamber indicates the intensity of any ionizing radiation.

Let us examine this process in greater detail. Energy furnished by the x-rays ejects an electron from one of the molecules in the gas in the chamber. The molecule thereupon becomes a positive ion and is pulled by the electric field towards the negative electrode. The ejected electron is accelerated in the opposite direction. After traveling a certain distance d in this electric field, the electron has acquired enough kinetic energy to knock another electron out of the next molecule it collides with. There are now two electrons where previously there was but one. Both electrons then start out on another energy-acquiring rush which culminates, after covering an additional distance d, in two more ionizing collisions, resulting in a new total of four electrons, and so on. This process is called impact ionization. The avalanche proceeds, with the total number of electrons doubling every time the front of the avalanche advances by a distance d. We should interject here that the ionization distance d is not a quantity applying precisely to each and every electron, but is rather a statistical mean taken for a large number of electrons. Eventually this snowballing cascade reaches the positive electrode where it delivers its charge to the metal and becomes simply a congregation of neutral molecules once more. In this way the charge of the initial electron, much too small to register as a current in the circuit, may be multiplied billions of times before it is collected. The amount of the multiplication depends on the composition of the gas, the size of the chamber, and the voltage between the plates.

The mathematics of this process goes as follows. Suppose each free electron in the chamber causes the production of n additional electrons by collision in traveling a unit distance along the electric field. You will recognize that n is the reciprocal of the ionization distance d mentioned in the preceding paragraph. The number dN of new electrons produced as the avalanche proceeds by a distance dx will be proportional to the number N of electrons already present. That is

$$dN/dx \propto N \qquad \text{or} \qquad dN/dx = nN.$$

We solve this to obtain

$$\boxed{N = N_0 e^{nx},}$$

where N_0 is the number of electrons present as the avalanche crosses the plane at $x = 0$ from which we start measuring distance. If we assign $x = 0$ to the place where the first electron is produced, then $N = e^{nx}$ merely gives the multiplication of the avalanche from that point to the electrode.

Ideally this action leads to a truly exponential growth of the avalanche with distance. In practice, however, there are some physical processes which cause deviation. The positively ionized gas molecules left behind as the avalanche roars by are thousands of times more massive than the electrons. Consequently their motion towards the negative electrode is relatively sluggish. The positive ion cloud has made but little progress by the time the electron avalanche has reached the positive electrode and discharged. The ion cloud introduces a space charge which partially shields the charge on the negative electrode, thus reducing the strength of the electric field which causes motion of the electron avalanche. As a consequence the specific ionization of the electrons (the number of ion pairs formed per unit distance) is reduced. Further, if the avalanche has a long way to go before reaching the electrode, it may grow to such a size that the number of positive ions it produces in a small volume dV begins to become an appreciable fraction of the number of molecules in that volume. This effect will further reduce the specific ionization.

The positive ions in their transport towards the negative electrode do not ordinarily produce further collision ionization. They move slowly, and the kinetic energy they acquire from the electric field is continually drained away by the jostling of neighboring molecules. The voltages used in ion-chamber operations are adjusted to ensure this condition. However, if the field strength is increased sufficiently, the positive ions themselves will produce new ion–electron pairs by collision. If the field strength is great enough to provide that a positive ion will produce, on average, more than one additional positive ion, then the avalanche grows and the gas in the chamber breaks down into an electric arc.

2.6 General Remarks

All exponential growth processes ultimately lead to infinite output unless something happens to supervene; and something always does. In the real world nothing can go to infinity. When the snowballing savings account grows large enough to drain away too much of the bank's resources in interest payments, the bank may farm out some of the account to other banks. When they, too, become overcommitted, they may simply terminate the account. A growing biological population is always limited ultimately by the food supply rate, if not sooner by some other agency. For example, the lemmings of Scandinavia limit their population by periodic orgies of mass suicide. Other animal populations, *Homo sapiens* to the contrary, limit their size by cutting back the reproduction rate or waging a war of acquisition when food becomes scarce or when crowding becomes acute. The ion avalanche either disappears when it discharges at the electrode or else throws the ion chamber into an arc condition having entirely different characteristics. The gossip explosion eventually runs out of people who haven't heard it already, and the chain letter scheme ultimately runs out of new suckers.

We see that exponential growth is a phenomenon which pervades many areas of science and human activity. For three typical cases we have derived these similar looking equations for growth:

$$
\begin{array}{ll}
\text{savings account} & \$ = \$_0 e^{rt}, \\[4pt]
\text{population} & N = N_0 e^{gt}, \\[4pt]
\text{ion avalanche} & N = N_0 e^{nx}.
\end{array}
\qquad (2.4)
$$

These expressions are similar because they all describe a particular type of phenomenon, regardless of the area in which it operates, be it finance, biology, or physics. Exponential growth is characteristic of the particular class of growth processes in which the rate of growth at any time is proportional to the amount of whatever it is that's growing.

In recent years concern has again arisen over the rate of increase of world population.† Unless it can be stabilized at a

---

† Such concern is of a recurrent nature. In 1798 the British economist Thomas Robert Malthus predicted that, since population growth was outpacing the food supply, growth, wars, pestilences, and starvation would supervene to kill off the excess population. Nineteenth- and twentieth-century improvements in agriculture have so far stalled off Malthus's Götterdämmerung, but what of the future?

level compatible with the Earth's capacity for producing food, catastrophy is inevitable. In some underdeveloped countries, notably India, nature has intervened in her own inexorable way to maintain balance between population and food supply through starvation. Other countries, for example Japan, have sought to improve the balance by waging wars of territorial expansion. Are these dismal alternatives to be the future prospect for mankind? We close this chapter with an adaptation of the epigram at its opening: The greatest shortcoming of the human race is man's inability or unwillingness to reckon with the exponential function.

▶ **Exercises**

2.1   What rate of compound interest would you require from an investment in order to double your capital in 10 years?

2.2   What is the doubling time for growth rates of (a) 1 %, (b) 6 %, and (c) 10 % per year?

2.3   Show that a characteristic of the exponential fuction $N = N_0 e^{kt}$ is that the time t_c for N to grow by any factor c remains constant throughout the growth.

2.4   If the population of a town grows steadily at the rate of 5 % per year, by what factor will such services as electric power, water supply, and sewage treatment have to be increased in one lifetime (70 yr)? Assume that each individual in the population consumes these services at a constant rate.

2.5   Repeat Exercise 2.4, this time assuming that the consumption of services by each individual increases at the rate of 5 % per year.

2.6   Show that the number N of kilometers of highways in the country can be expected to grow exponentially if these assumptions hold: (a) the number D of kilometers driven per year by the population is proportional to N; (b) highway revenue $R = bN$, where b is the tax per car-kilometer; and (c) a constant fraction of R is used for construction of new highways.

2.7   The US per capita energy consumption, in terms of equivalent tons of coal per person per year, is given in the following table:

Year	1912	1925	1937	1950	1962	1975
Tons	1.00	1.68	3.25	5.28	7.10	11.33

Is this growth exponential or nearly so? Approximately how much coal equivalent per capita will be required in the year 2000 to maintain the growth if it continues at the present rate?

2.8   A nuclear chain reaction 'goes critical' whenever conditions ensure that the disintegration of one nucleus triggers the subsequent disintegration, on average, of more than one additional nucleus. Show that this condition leads to an exponential growth of the explosion.

2.9   A certain chemical reaction proceeds at a rate dependent on temperature according to the equation: rate $= a + bT$, where a and b are constants. The reaction generates heat and increases the temperature of the reactants at a rate proportional to the reaction rate. Show that this situation leads to an exponential increase of temperature with time.

2.10   Show that a convenient relation for finding the approximate time required for something experiencing exponential growth to double in amount is:

$$\text{doubling time} = \frac{70}{\text{percentage growth per unit time}}.$$

2.11   Replot figure 2.3 using the $(1 + r)^n$ formula instead of the e^{nr} formula. Show that this plot is also exponential, with a very slightly different slope.

2.12   A snowball rolling down an incline at constant linear speed picks up snow to the thickness of the snowfall in each rotation. Why is the increase of the mass of the snowball *not* an exponential function of time?

2.13   A poorly adjusted public address system sometimes breaks into a howl because sound from the loudspeaker, picked up by the microphone and amplified, causes the loudspeaker output to become louder. Does this kind of feedback lead to an exponential increase of sound intensity? What prevents this increase from going to infinity? What factors determine the *rate* of the build-up?

2.14   The population of bacteria in a culture flask has a doubling time of one day. If the population is one million on a particular day, (a) when was it half a million and (b) when will it be sixteen million?

3 Exponential Decay with Time

Analogies are useful for analysis in unexplored fields. By means of analogies an unfamiliar system may be compared with one that is better known. The relations and actions are more easily visualized, the mathematics more readily applied, and the analytical solutions more readily obtained in the familiar system.

Harry F Olson

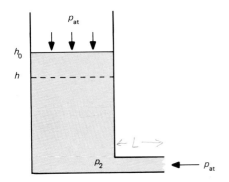

Figure 3.1 The water tank is emptying through the outflow pipe.

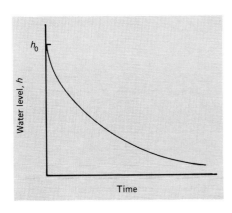

Figure 3.2 As time goes on the water level drops, but at an ever-decreasing rate.

sectional area A, then $V = Ah$ and $dV/dt = A(dh/dt)$. Hence,

$$\frac{dh}{dt} = -\frac{F\rho gh}{A}.$$

We can rewrite this equation in such a form that it can be integrated,

$$\frac{dh}{h} = -\frac{F\rho g}{A}dt.$$

Solving this differential equation we obtain

$$\ln h = -\frac{F\rho g}{A}t + c,$$

where c is the constant of integration. Taking antilogarithms,

$$h = e^{-(F\rho g/A)t}e^c.$$

Since c is a constant yet to be determined, e^c is also a constant; for the time being call it q. Then we have

$$h = qe^{-(F\rho g/A)t}.$$

There remains only the task of determining the value of q from the conditions of the problem. At the very start of the outflow $t = 0$ and $h = h_0$. At that instant the factor $e^{-(F\rho g/A)t} = e^c = 1$, from which we obtain $h_0 = q$ and finally

$$h = h_0 e^{-(F\rho g/A)t}. \tag{3.2}$$

We see, then, that the shape of the decay curve of figure 3.2 corresponds to an *exponential* drop of the water level in the tank. This exponential type of decay is typical of many physical processes in nature. We shall presently examine some more of them. Any process in which the *rate* of decrease of something is directly proportional to the *amount* of that something remaining at any instant inevitably leads to an exponential type of decay.

Exponential decay is so widespread a phenomenon that it is worthwhile here to pause long enough to study some of its properties. First of all, the time scale of the decay curve depends on all the constant factors in the exponent, that is upon $F\rho g/A$. We can hasten the decay process by increasing any of the factors in the numerator or by decreasing the denominator. We can slow down the decay by changing any of the factors in the other direction.

Suppose $L \to 0$ so that $F \to \infty$. This means $h \to 0$ in very short times which is a nonsense!

$\frac{dV}{dt}$ can not be ∞!

Something is missing.

(Escape velocity from the outlet is $v = \sqrt{2gh}$ when $L = 0$)

A special significance is attached to the time at which $t = A/F\rho g$. At that instant the argument of the exponent is equal to unity and $h = h_0 e^{-1} = (1/e)h_0$. At that particular instant the water level has sunk to $1/e$ times (or approximately 37% of) its original height, and the time of this instant is called the one-over-e time, the relaxation time, or the *time constant* of the process. Observe that equal time intervals do not produce equal reductions of the water level. At the instant of two time constants, for example, the water level has not sunk to $1/2e$ times its initial height, but rather to $(1/e)^2$ times its initial height. Thus equal time intervals produce equal reduction *factors*. Each additional time increment of one time constant produces a further reduction of height by another factor of $1/e$, no matter where on the curve you choose to start counting time.

Observe that according to equation (3.2) the water level approaches, but never actually reaches, zero height. By letting the process continue long enough, however, we can make the water level come arbitrarily as close to zero as we wish. For example, if we wait for five time constants, the water level will sink to $h_0 (1/e)^5 = 0.007h_0$ or a little less than 1% of its initial height. We might then say that the tank was empty for most intents and purposes, although we do so with the mental reservation that it will never become completely empty.

3.2 Discharging of a Capacitor

Turn now to the examination of another situation leading to an exponential type of decay—the discharging of an electrical capacitor through a series resistor. In the circuit of figure 3.3 we have a capacitor of capacitance C in series with a switch s_2 and a resistor of resistance R. The upper plate of the capacitor can be charged to the potential V_0 by momentarily closing the switch s_1 in the auxiliary circuit shown in the dashed box. With the capacitor then initially charged to potential difference V_0 we can open switch s_1 and disregard the charging circuit in all that follows. We now close switch s_2, simultaneously start a stopwatch, and let the capacitor discharge through the discharge circuit containing s_2 and the resistor. We wish to know how the potential V on the upper plate of the capacitor will decrease with time after the closing of s_2.

As before, let's think our way through the discharging process to a statement of what we expect to happen. Charge from the upper plate will flow as a current through the resistor around to the lower plate. The *rate* of the flow is given by Ohm's law,

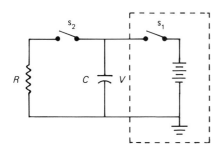

Figure 3.3 Once it is charged, the capacitor in this circuit discharges exponentially.

$I = V/R$ (according to equation 1.2), where V is the instantaneous potential on the upper plate of the capacitor. As the flow continues and charge is transferred from the upper to the lower plate, the potential V on the upper plate becomes smaller and smaller, and with a decreasing V the current decreases too. The discharge thus proceeds at an ever-decreasing rate. Here we have another situation in which the rate of the process at any instant is proportional to the quantity of something remaining at that instant. We expect, therefore, that the decay of V will be exponential in nature. Let's see.

From equation 1.2,

$$\frac{dQ}{dt} = -\frac{(V-0)}{R} = -\frac{V}{R}.$$

Since the charge Q and voltage V of a capacitor are related to the capacitance C by $Q = CV$, we can rewrite the equation above as

$$\frac{dQ}{dt} = C\frac{dV}{dt} = -\frac{V}{R} \qquad \text{or} \qquad \frac{dV}{V} = -\frac{dt}{RC}.$$

Solving this differential equation gives,

$$\ln V = \frac{-t}{RC} + \text{constant},$$

and taking antilogarithms we find

$$V = \text{constant} \times e^{-t/RC}.$$

Evaluating the constant of integration from the condition that at time $t = 0$, $V = V_0$, we obtain finally

$$\boxed{V = V_0 e^{-t/RC}.} \tag{3.3}$$

So the potential difference between the capacitor plates decreases exponentially with time. The decrease is characterized by a time constant equal to the product RC. As a practical example, a $1\,\mu F$ capacitor discharging through a $1\,M\Omega$ resistor will have a time constant of 1s. That is, after 1s the potential difference between the capacitor plates will have decayed to about 37% of its initial value.

▶ **Example 3.1** An electrostatic voltmeter has a range of 0 to 250 V and the smallest potential difference that can be estimated with it is 0.5 V. A $1.0\,\mu F$ capacitor is charged to 200 V and then allowed to discharge through a $1.0\,M\Omega$ resistor. In what time will the capacitor be completely discharged as indicated by the voltmeter?

From equation (3.3) we find:

$$e^{-t/RC} = V/V_0,$$
$$e^{t/RC} = V_0/V,$$
$$t = RC \ln (V_0/V).$$

In this example $R = 1.0 \times 10^6 \, \Omega$, $C = 1.0 \times 10^{-6} \, \text{F}$, $V_0 = 200 \, \text{V}$, $V = 0.5 \, \text{V}$. Therefore

$$t = (1.0 \times 10^6)(1.0 \times 10^{-6}) \ln (200/0.5) = 6.0 \, \text{s}.$$

Since the decay equation for the capacitor is similar in form to that for the water tank, it may be instructive to write them, one under the other, to see if we can identify the analogous features of the two physical processes:

water tank $h = h_0 e^{-(F\rho g/A)t}$,

capacitor $V = V_0 e^{-(1/RC)t}$.

Clearly, the instantaneous potential difference between the capacitor plates corresponds to the height of the water in the tank, and the resistance R of the electrical discharge circuit corresponds to the flow resistance $1/F$ of the outflow pipe from the tank. It may be less obvious that the capacitance C of the capacitor corresponds to the combination of factors $A/\rho g$ for the tank.

3.3 Radioactive Decay

Consider 1 mg of pure radium metal, lying on a table top. This material, as you know, is one of the naturally radioactive substances: the nuclei of its atoms are unstable and spontaneously disintegrate. A radium nucleus disintegrates by throwing out an alpha particle and transforming itself into the nucleus of an atom of radon. The radon nuclei are not permanently stable but in turn suffer a series of further disintegrations, becoming eventually nuclei of one of the isotopes of lead.

It is impossible to predict which of a collection of radium nuclei are to be the next ones to go, but we do know statistically that any individual radium nucleus has a probability $p = 1.37 \times 10^{-11}$ of going within the next second. If we have an initial large number N of undisintegrated radium nuclei in our sample at any instant, then the number dN/dt of those which will disintegrate within the next second is given by the product of their number and the probability that any one will disintegrate in the next second, that is $dN/dt = -pN$.

Here is another situation in which the *rate* dN/dt of a decay process is directly proportional to the *amount* N of the reactant in existence at any particular instant. As the disintegration continues, the remaining number of radium nuclei decreases, while the rate of further decrease becomes smaller and smaller. Clearly this is another example of an exponential decay process.

The solution of the radioactive decay equation is:

$$N = N_0 e^{-pt}, \tag{3.4}$$

where N_0 is the original number of radium nuclei. The time constant of the process, that is the time required for the number of unexploded radium nuclei in the sample to decrease to $1/e$ times the original starting number, is $1/p = 1/(1.37 \times 10^{-11}) = 7 \times 10^{10}$ s, which is around 2200 yr. The physicist has his own term for this time constant; he calls it the 'average life' because of a particular property of the exponential decay. The average life of a starting population of N_0 nuclei is given by the sum of the lives of all its members, divided by the starting number. That is,

$$\text{average life} = \frac{1}{N_0} \int_0^\infty N \, dt = \frac{1}{N_0} \int_0^\infty N_0 e^{-pt} \, dt$$

$$= \int_0^\infty e^{-pt} \, dt = \frac{1}{p}.$$

In the literature of radioactivity, the quantity we have called decay probability p is often designated the decay constant λ.

The time required for the number N of radioactive atoms in a pure radioactive sample to decay to half the original number N_0 is called its half-life $t_{\frac{1}{2}}$. To determine this half-life we note, from equation (3.4), that $N = \frac{1}{2}N_0$ when $t = t_{\frac{1}{2}}$. Thus,

$$\tfrac{1}{2}N_0 = N_0 e^{-\lambda t_{\frac{1}{2}}}, \quad \text{or} \quad \ln \tfrac{1}{2} = -\lambda t_{\frac{1}{2}}.$$

Hence,

$$t_{\frac{1}{2}} = -\frac{\ln \frac{1}{2}}{\lambda} = \frac{\ln 2}{\lambda} = \frac{0.693}{\lambda}.$$

▶ **Example 3.2** It is found that the activity of a sample of an artificially produced radioactive isotope decreases from 0.010 to 0.003 Ci[†] in 60 days. What is the half-life of this radio-isotope?

† The curie (Ci) is not an SI unit, but is still used occasionally. 1 Ci is equivalent to 3.7×10^{10} Bq (1 Bq = 1 disintegration/s).

From equation (3.4) we find that

$$\frac{N}{N_0} = e^{-\lambda t},$$

$$\ln \frac{N_0}{N} = \lambda t,$$

$$\lambda = \frac{1}{t} \ln \frac{N_0}{N}.$$

But, since activity A is proportional to the number N of radioactive atoms,

$$\frac{A_0}{A} = \frac{N_0}{N},$$

and therefore

$$\lambda = \frac{1}{t} \ln \frac{A_0}{A}.$$

In this example $A = 0.003\,\text{Ci}$, $A_0 = 0.010\,\text{Ci}$ and $t = 60$ days. Substituting in the above equation gives

$$\lambda = \frac{1}{60} \ln \frac{0.010}{0.003},$$

$$= 0.020.$$

From above the half-life $t_{\frac{1}{2}} = \ln 2 / \lambda$, therefore

$$t_{\frac{1}{2}} = 0.693/0.020,$$

$$= 34.65 \text{ days}.$$

3.4 Cooling a Warm Object

It is part of everyone's experience that a warm object left by itself in cooler surroundings eventually cools to the temperature of its surroundings. This observation was formalized by Newton who stated that the *rate* of decrease of the temperature of the object is approximately proportional to the difference in temperature between the object and its environment, if this difference is not too large to start with. This rate includes heat losses by all mechanisms, whereas the more precise statement of Fourier's equation (equation 1.4) is applicable to cases of conduction only.

Let us consider, then, an object initially at a somewhat elevated temperature T_1, placed in an environment at T_0 degrees. How fast will it cool, and what will its temperature be at any subsequent time? We have $dT/dt \simeq -(T - T_0)$, or $dT/dt = -B(T - T_0)$, where B is the constant of proportionality in s^{-1}. B will be given by some complicated function of the size, shape, specific heat, and surface condition of the object. It will involve also the nature and convective properties of the gas the object is immersed in and the thermal conductivity of the support the object is placed upon. By comparison of this differential equation with those of the preceding cases of the water tank and the discharging electrical capacitor, we can write the solution as

$$T - T_0 = (T_1 - T_0)e^{-Bt}.$$ (3.5)

This equation is that of a typical exponential decay with a time constant of $1/B$.

▶ **Example 3.3** Show that Newton's law of cooling is an approximation (for small temperature changes) of Stefan's law: the rate of radiation from a hot body is proportional to the *fourth* power of its absolute temperature.

The net rate of loss of energy dQ/dt by an ideal radiator at temperature T on cooling in an environment at temperature T_0 is the difference between its rate of loss of energy and its rate of absorption of energy from its surroundings. Implementing Stefan's law gives:

$$\frac{dQ}{dt} = \sigma(T^4 - T_0^4),$$

where σ is Stefan's constant. But when a body loses heat, the rate of fall of temperature is proportional to the rate of loss of energy:

$$\frac{dT}{dt} = \frac{1}{mc}\frac{dQ}{dt},$$

where m is the mass of the body and c is its specific heat capacity. Substituting for dQ/dt in this equation gives

$$\frac{dT}{dt} = \frac{\sigma}{mc}(T^4 - T_0^4),$$

$$= \frac{\sigma}{mc}(T - T_0)(T + T_0)(T^2 + T_0^2).$$

As $T \to T_0$, the second and third parenthetical factors on the right-hand side become essentially constant, while the factor $T - T_0$ (though small) continues to change. Thus,

$$\frac{dT}{dt} \propto T - T_0.$$

Stefan's law considers cooling by radiation alone in an evacuated chamber. Of more practical interest, especially in calorimetry, is the case of an object cooling by convection, conduction, and radiation losses. Interestingly, Newton's law of cooling has a much greater range of application in this more complex case! For $T = 300\,\text{K}$ and $T_0 = 50\,\text{K}$, the prediction of equation (3.5) above may be too low by some 20 % when applied to a body losing energy by radiation alone, but low by less than 1 % when applied to a body cooling by convection and radiation.

3.5 Summary

This chapter can be summed up by pointing out that (a) an exponential decline is characteristic of any decay process where the *rate* is proportional to the amount of the decaying quantity remaining at any instant, and (b) such processes are describable by an exponential decay factor $e^{-t/\tau}$ where τ is a quantity or group of quantities appropriate to the particular situation and having the physical meaning of a time constant.

Perhaps as you pondered these developments you have observed that exponential decay is describable by the same form of mathematical expression as that derived in Chapter 2 to describe exponential growth, except that the exponent is negative. In exponential growth the quantity doing the growing increases toward infinity. In exponential decay the quantity considered decreases toward zero. Whether something grows or declines depends on the sign of the characterizing rate factor. Whether a population grows or withers away depends on whether the birth rate exceeds the death rate, leading to a positive exponent in the rate equation, or whether the death rate exceeds the birth rate leading to a negative exponent.

The control of industrial operations is often exercised by controling the sign of the appropriate rate quantity. For example, in the core of a nuclear reactor, the reaction 'goes critical' when a neutron released in one fission causes, on average, one additional fission. A critical reaction has the same number of fissions in each succeeding generation. If the number of fissions increases from one generation to the next, the power is increasing and the reaction is said to be supercritical. By the insertion of neutron-absorbing control rods into the core the operator can regulate the growth, steady maintenance, or decay of the reaction, depending on whether the regeneration factor exceeds, equals, or falls short of unity. In the former case the exponent of the rate equation is positive and the reaction grows, developing increasing power; in the latter case the exponent is negative, the reaction decays toward zero, and the power output declines.

▶ **Exercises**

3.1   When a ^{14}C atom is formed in the atmosphere, it reacts with oxygen to form a CO_2 molecule and may enter a living organism. After the organism dies, the ^{14}C already in it decays with a half-life of 5700 yr. If a sample of carbon from an excavation of an ancient site gives 18 counts/min as its activity and an equal mass of freshly charred wood gives 52 counts/min, what is the age of the carbon found at the site?

3.2   When a police inspector found a body at 2 AM its temperature was 81°F; by 5 AM it had fallen to 76°F. Assuming that the room temperature remained constant at 50°F, estimate the time of death (when the body temperature was 98.5°F).

3.3   Radium atoms are unstable; they emit alpha particles as they disintegrate at a rate of about 0.045 per cent per year. Show that the number of alpha particles emitted per gram of radium in any one day is about 3.3×10^{15}. (Atomic weight Ra = 226; Avogadro's number = 6.06 $\times 10^{23}$.)

3.4   An automobile with its motor out of gear is coasting along a level road. It encounters an opposing frictional force in the tires, bearings, and air flow which is proportional to the velocity. Derive an expression for the velocity of the car as a function of time.

3.5   Atmospheric pressure, p_0 at sea level, decreases vertically upward. Assuming that the ideal gas law holds and that the temperature is constant, show that in each thin stratum (where the air has density ρ) $dp = -\rho dh$ and derive Halley's law for pressure p at height h: $p = p_0 e^{-\rho_0 h/p_0}$. Assume air to be a perfect gas.

3.6   A man dies and leaves his widow a sum of money in a savings account earning 5% interest per year. Each year she withdraws 6% of the remaining amount for living expenses. Calculate the value of her nest egg at the end of 10 years if its purchasing power is further eroded by inflation at the rate of 8% per year.

3.7   Assume that 600 dice are thrown on a table and that all which show a six are removed. The remainder are shaken and again spread on the table, the sixes removed, and the process continued. Show that the 'half-life' n of a die in this process is about four throws. HINT: Justify finding n from $(5/6)^n = \frac{1}{2}$.

3.8   A party of castaway explorers agree that they will ration their non-replenishable stock of food at the rate of 5% of the remaining stock each day. After one week, becoming hungry, they decide to increase their rate of consumption to 10% of the remaining stock each day. What fraction of the original cache will remain one week later?

3.9   After sunset the positive and negative ions produced by ultraviolet light in the ionosphere begin to recombine. The recombination rate is proportional to the number of ions of one kind present per cubic meter at any time, times the number of ions of the other kind remaining for them to recombine with; that is, to the square of the number of ions of either kind. Derive the equation for ion density as a function of time. Is it exponential?

3.10   A leaky 2.0 μF capacitor is given an initial charge and then isolated. The potential difference between its plates drops from V_0 to $\frac{1}{4}V_0$ in 3.0 s. Find the equivalent resistance between the capacitor plates.

3.11   A large snowball, initially at 0 °C throughout, melts at a rate proportional to the area of its surface exposed to warm air. Does the unmelted mass decrease exponentially with time?

3.12   Suppose the water tank of figure 3.1 to be conical rather than cylindrical, with the apex at the bottom and the outflow pipe coming out at this point. What modifications will you have to introduce into the derivation of equation (3.2)? Will the resulting expression be exponential?

3.13   Suppose that the dielectric between the plates of the capacitor of figure 3.3 is highly compressible so that the electrostatic attraction between the plates causes a decrease in their separation by an amount proportional to the voltage between them. What effect will this perturbation have on the shape of the discharging curve?

3.14   In days when food was kept cool in ice boxes rather than in electric refrigerators, a frugal householder might place newspapers over the ice. How did this slow its melting? Were there any disadvantages?

3.15   It is desired to determine the mass m of a minute quantity of a radioactive isotope which is an alpha-particle emitter with a known decay constant λ. If this radioactive material is placed between the plates of the capacitor shown in figure 3.3 the rate at which the capacitor discharges (as measured with an electrostatic voltmeter) will be increased. Why? Describe an experiment by which you could use this fact to determine m.

4 Exponential Decay with Distance

In every department of physical science there is only so much science, properly so-called, as there is mathematics.

Immanuel Kant

4.1 Introduction

The illumination received from a bare light bulb decreases as you increase your distance from it. This decrease is largely a consequence of geometry. At a distance of 1 m from the bulb, its radiation sends energy through a spherical surface of area $4\pi(1)^2 \, \text{m}^2$. At a distance of 2 m, the same energy is spread over an area $4\pi(2)^2 \, \text{m}^2$, which is four times as large. So doubling the distance reduces the illumination to one fourth, i.e. the illumination from a point source varies inversely as the square of the distance from the source.

This sort of attenuation due to the *spreading* of a spherical wave (with no loss in total energy) is *not* the subject of this chapter. Rather we shall be concerned with the decrease in an essentially one-directional flow due to the *absorption* or removal of energy at each element of path in that flow. An example of a decay of this type is the diminution of the intensity of a beam of light as it progresses through a uniform absorbing medium. The farther the beam penetrates the less intense it becomes because of the absorption of energy from the beam by the molecular processes in the medium. This type of flow accompanied by absorption of energy leads to an exponential decay with distance.

4.2 Absorption of Radiation

There are many situations where we are interested in the absorption of radiation: in photographic light filters, in protective lead screens around x-ray tubes, or in the atmosphere. As a beam of radiation passes through an absorbing medium, in each unit distance of its path it suffers a loss of intensity which is proportional to the intensity at the beginning of that unit of path. If the beam is traveling in the x direction,

$$\frac{\text{d}I}{\text{d}x} = -kI,$$

where k is the constant of proportionality called the absorption coefficient. The solution of this equation is

$$\boxed{I = I_0 e^{-kx},} \qquad (4.1)$$

where I_0 is the initial intensity of the beam as it enters the absorbing medium. The decline in intensity is exponential with distance in the medium.

As you know, quantum theory describes radiation as having particle properties; energy is transferred in discrete packets: photons or quanta. Yet in discussing absorption we have treated I as a mathematically continuous variable. Similarly, in radioactivity we dealt with discrete nuclei, yet we treated N as a continuous variable. Our mathematical treatment is justifiable in both cases provided we are dealing with a statistically large number of nuclei or photons.

► **Example 4.1** A certain red filter glass has an absorption coefficient of $0.20\,\text{mm}^{-1}$ for light of wavelength $0.65\,\mu\text{m}$. What thickness x is needed to reduce the intensity of the red beam to 0.37 times its original intensity?

From equation (4.1) we have:

$$\frac{I}{I_0} = e^{-kx},$$

$$x = -\frac{1}{k}\ln\frac{I}{I_0}.$$

In this example $I/I_0 = 0.37$ and $k = 0.20\,\text{mm}^{-1}$. Substituting in the above equation gives:

$$x = -\frac{1}{0.20}\ln 0.37,$$
$$= -5(-0.99),$$
$$= 5.0\,\text{mm}.$$

When the absorbing medium is a solution of an absorbing solute in an otherwise clear liquid, Beer's law states

$$I = I_0\, e^{-c\beta x}, \tag{4.2}$$

where c is the solute concentration, β its specific absorptivity, and x the path length through the solution. An absorbing material can thus be characterized by its specific absorptivity, a constant regardless of its concentration in the solution.

4.3 Heat Conduction along a Bare Rod

In §1.3, we studied the flow of heat along a thermally conducting rod whose sides were insulated so that heat could not escape laterally. We are now going to consider a different arrangement in which the lateral surfaces are bare and heat can leak out into the

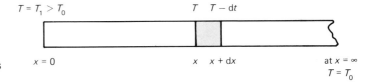

Figure 4.1 The temperature decreases from left to right along the rod.

surroundings. Consider a long, cylindrical conducting rod (figure 4.1) whose left-hand end is maintained at some elevated temperature T_1 and whose right-hand end is so far away that it is essentially at room temperature, T_0. Heat escapes from each lateral surface element at a rate given by Newton's law as proportional to the difference between the temperature at the element and the temperature of the surroundings. There will be a progressive decrease in temperature from left to right along the rod because of this heat loss. If the rod and its surfaces are uniform, the decrease in temperature per unit distance will be proportional to the difference between the local temperature of the rod and room temperature. Thus,

$$-\frac{dT}{dx} = L(T - T_0),$$

where L is a loss-rate factor depending on the surface area of the rod per unit length, the dissipative characteristics of the surface, the thermal conductivity of the rod, and the conductive and convective properties of the surrounding air. The solution of this equation is

$$T - T_0 = (T_1 - T_0)e^{-Lx}. \tag{4.3}$$

The decrease in temperature along the rod is exponential in nature.

4.4 Two-wire Electrical Transmission Line

Consider a long transmission line (figure 4.2) to be made up of a large number of sections in series, each section being one unit

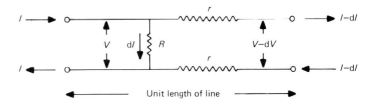

Figure 4.2 A transmission line is made up of a large number of unit sections in series.

distance in length in the x direction. The series resistance r per unit length of each conductor might be $1\,\text{m}\Omega\,\text{m}^{-1}$. There exists a leakage resistance R between the wires, because of the fact that there is no insulation with infinite resistance. Resistance R might be $1\,\text{M}\Omega$ in each $1\,\text{m}$ section. There is a positive current I into the section from left to right in the upper wire, and an equal current I out of the section from right to left in the lower wire. Let the voltage between the wires be V at the left-hand end of the section. The current leaving the section in the upper wire at the right-hand end is less than the current entering at the left by the amount of leakage current between the wires within the section. This leakage current is a function of the voltage between the wires at the particular section we are examining. Thus

$$\frac{dI}{dx} = -\frac{V}{R}. \tag{4.4}$$

The minus sign indicates that as one goes to the right (x increasing) the current decreases. Also, there is a voltage drop along the wires because of the Ir drop of the current in the wire resistances. This drop depends on the line current in that particular section, thus

$$\frac{dV}{dx} = -2Ir. \tag{4.5}$$

Differentiating equation (4.5) with respect to x, and substituting for dI/dx from equation (4.4) we obtain:

$$\frac{d^2V}{dx^2} = \frac{2r}{R}V.$$

The complete solution of this second-order linear differential equation is

$$V = c_1 e^{x\sqrt{(2r/R)}} + c_2 e^{-x\sqrt{(2r/R)}},$$

where c_1 and c_2 are the constants of integration.

The first term of this *mathematical* solution is inappropriate to the *physical* solution of the transmission line, since it predicts a line voltage which approaches infinity as the line is made longer and longer. We know this cannot be so, and therefore the meaningful solution for this particular situation consists solely of the second term on the right-hand side of the equation, giving

$$V = c_2 e^{-x\sqrt{(2r/R)}}.$$

To evaluate the constant of integration, suppose we specify that the input voltage applied to the line at the generator station, located at $x = 0$, is V_0. Then $c_2 = V_0$ and finally

$$V = V_0 e^{-x\sqrt{(2r/R)}}. \qquad (4.6)$$

We see from this equation that the voltage between the wires decreases exponentially with distance as we go along the line away from the generating station. The one-over-e distance of this decrease is $\sqrt{(R/2r)}$.

It must be stated that the analysis given here is valid only when the transmission line is sufficiently long that the voltage at the far end has decayed essentially to zero. The only input current, then, is that required to make up the leakage losses over the entire length of the line. Such a line would obviously have no utility, as it would deliver no current at the far end. In practice we don't make transmission lines so long that the leakage losses amount to more than a few per cent of the input. The usual practice is to design the line so that its one-over-e length is several times the actual distance to be covered. The terminating load is then designed to operate on the voltage prevailing on the line at the point of termination and to draw just as much current as the rest of the line would draw if it were infinitely long.

If the load is removed altogether, and the line left open circuited at its termination, the exponential decay of the voltage along the line will be less than that calculated from equation (4.6). If the load draws more than the normal current, there will be an additional Ir drop in each line section, and the decay will be steepened. Figure 4.3 illustrates these variations.

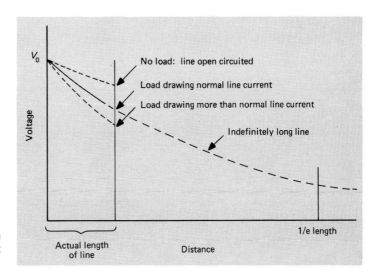

Figure 4.3 The voltage decreases exponentially along the line, depending on the leakage parameter and on the current drawn by the terminating load.

▶ **Example 4.2** A certain very long transmission line has a series resistance of $0.01\,\Omega\,\text{km}^{-1}$ and a shunt leakage conductance $1/R$ of $0.001\,\Omega^{-1}\,\text{km}$. What fraction of the input voltage will the reduced voltage be 50 km down the line?

From equation (4.6) we have

$$\frac{V}{V_0} = e^{-x\sqrt{(2r/R)}}.$$

In this example $x = 50\,\text{km}$, $r = 0.01\,\Omega\,\text{km}^{-1}$, $1/R = 0.001\,\Omega^{-1}\,\text{km}$, therefore

$$\frac{V}{V_0} = e^{-50\sqrt{[2(0.01)(0.001)]}},$$

$$= e^{-0.224},$$

$$= 0.80,$$

i.e. the reduced voltage 50 km down the line is 0.80 times the input voltage, V_0.

To find out how the *current* decreases as we go along the line we differentiate equation (4.6), obtaining

$$\frac{dV}{dx} = -\sqrt{(2r/R)}.V_0 e^{-x\sqrt{(2r/R)}}.$$

Substituting this derivative in equation (4.5) we have

$$\frac{dV}{dx} = -2Ir = -\sqrt{(2r/R)}.V_0 e^{-x\sqrt{(2r/R)}},$$

from which

$$I = \sqrt{(1/2rR)}.V_0 e^{-x\sqrt{(2r/R)}}. \tag{4.7}$$

If we specify that the current at the generator end of the line is I_0 when $x = 0$, then, by substitution in equation (4.7),

$$\boxed{I = I_0 e^{-x\sqrt{(2r/R)}}.} \tag{4.8}$$

We see from equation (4.8) that the current, too, decreases as we go away from the generating station, *with the same exponential decay factor as the voltage.*

In this decay factor, $e^{-x\sqrt{(2r/R)}}$, the constant quantity $\sqrt{(2r/R)}$ is determined solely by the properties of the transmission line and not by where we are along it or by the voltage and current being applied at the generator end. It is called by electrical engineers the *attenuation constant.* Another way of describing the 'lossiness' of a transmission line is to express the fractional decay per kilometer or mile in *nepers*. One neper is a loss by a factor of $1/e$. The expression 'so many nepers per mile' is commonly used.

While the analysis given here is appropriate for DC transmission lines, the same general results are also obtained for AC transmission lines. In the AC case, however, one must take account also of the series inductance and shunt capacitance of the wires. These matters will be considered in Chapter 18.

4.5 Electrical Analog for Heat Flow

The similarity between the equations describing the fall-off of temperature in a leaky thermal conductor and the fall-off of voltage on a leaky transmission line suggests that interesting simulations between thermal and electrical situations can be developed. Electrical measurements are usually easier to make than thermal measurements. Therefore, in laboratory investigations of heat flow it is often found convenient to make the actual measurements on a network of electrical resistors assembled to simulate the thermal situation.

Heat flow in a thermal system can be simulated by an electric current in an electrical system having a similar arrangement of elements. The voltage at different points in the electrical system will be in proportion to the temperatures at the corresponding points in the thermal system.

As a case in point, referring to figure 4.4, we can simulate the

Figure 4.4 The voltage at any point on the long resistance wire (*b*) is a simulation of the temperature at the corresponding point of the thermally conducting rod (*a*).

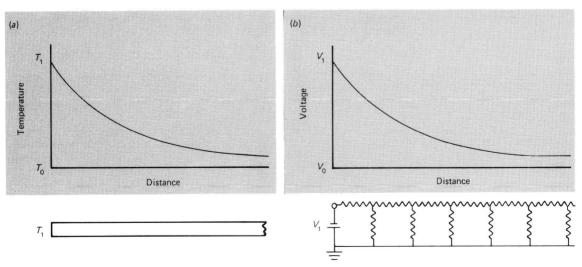

thermal system we presented in §4.3 (that of heat conduction along a rod with lateral heat loss) by studying the electrical currents and voltages at different points on a long resistance wire to which leakage conductances to ground connect at equal intervals. The progressive current losses through the leakage conductances correspond to the heat losses through the sides of the rod. The voltages at different points on the resistance wire will be proportional to the temperatures at the corresponding points on the rod. In assembling the simulating network one must ensure that the conductance per unit length of the resistance wire bears the same ratio to the leakage conductance that the thermal conductance of the rod bears to its lateral loss conductances. All elements of the simulation must have values proportional to those of the corresponding elements of the thermal situation being simulated.

4.6 Summary

From the examples we have presented, it should be clear that exponential decrease with distance occurs whenever there is a constant *fractional* decline of the decaying quantity per unit distance. That the decaying quantity may be a radiation intensity, a temperature, a current or voltage, a pressure, etc, should be no surprise. The purpose of this and the adjoining chapters dealing with exponential relationships in physics is to illustrate the widespread occurrence of such relationships and to show how the characterizing features of each exponential type can be identified.

► Exercises

4.1   A party of 200 settlers bound for California by wagon train suffers a brush with hostile Indians every 100 miles. At each encounter the Indians manage to kill 5 % of the surviving settlers. Show that only 94 of them will reach California 1500 miles away.

4.2   Water is flowing through a long hose with leaky sidewalls. Using the approach developed in § 4.3, derive an expression for the progressive drop in pressure along the hose. Assume the leaks are uniformly distributed over the length of the hose. Does the pressure decrease exponentially with distance?

4.3   The voltage at the input end of a long DC transmission line is 400 V. One kilometer down the line the voltage is 395 V. Show that the voltage at a point 10 km farther on is 348 V.

4.4   Assuming the atmosphere to be of homogeneous composition, obeying the ideal gas law with density proportional to pressure, show that the density decreases exponentially with altitude. The Earth's gravity is essentially constant throughout the height of the atmosphere.

4.5   Lead has an absorption coefficient of 0.65 cm^{-1} for hard x-rays. An x-ray tube is surrounded by a lead radiation shield 5 cm thick. What fraction of the generated x-rays penetrate the shield?

4.6   The specific absorptivity of copper ions for light of wavelength 5000 Å is 0.911 mole^{-1} cm^{-1}. Light of this wavelength passing through a glass cell 2 cm wide containing a solution of copper sulfate is observed to be reduced in intensity to 16% of the intensity of the same beam passing through the empty cell alone. What is the molar concentration of the copper sulfate?

4.7   Sound traveling through the inside of an acoustic pipe suffers loss of intensity because of the leakage of heat from regions of compression to adjacent regions of rarefaction, delivery of heat to and absorption of heat from the walls of the pipe, and molecular absorption. The overall rate of energy loss from the sound waves per unit length of pipe is proportional to the intensity. Assuming that there are no reflections from the far end of the pipe to complicate the situation, show that the resulting decay of sound intensity is exponential with distance along the pipe.

4.8   Hemoglobin has an absorption band at wavelength 4170 Å. Blood contained in a 1 cm thick optical cell is placed in a beam of light of this wavelength. The intensity of the emerging light is found to be one fourth that of the incident light. Find the linear absorption coefficient.

4.9   The corneum layer at the surface of a person's skin has an absorption coefficient of 315 cm^{-1} for ultraviolet light of wavelength 3200 Å. When sunlight of intensity 2 W m^{-2} and wavelength 3200 Å irradiates the skin, calculate the number of photons per second per square centimeter which reach the Malpighian layer containing pigment granules located 25 μm below the surface of the skin.

4.10   In Exercise 3.4 describing the coasting automobile, you were asked if the velocity – time relationship would be exponential. Now think through the question of whether the velocity–*distance* relationship should be exponential.

4.11   Boy Scouts on a hike agree that at the end of every mile they will take a rest. The first rest will be of 5 minutes' duration, the next 10 minutes', the next 15 minutes', and so on. Does their *average* speed decrease exponentially with distance? Assume that when they are walking their speed is 3 mph.

4.12   Suppose the thermally conducting rod of §4.3 had a linear taper from one end to the other. How would you modify the analysis to deal with this situation? Do you think that the decrease in temperature along the rod would be exponential?

4.13   The walls of a refrigerator consist of a three-layer structure, with layers having thicknesses of 1, 20, and 1 mm and thermal conductivities of 0.01, 0.0005, and 0.5 J/s/°C/m, respectively. The inside of the refrigerator is to be maintained at $-20\,°C$ and the outside at $+20\,°C$. Set up the electrical analog of this arrangement and determine the temperatures at the two inner-layer boundaries.

4.14   If the refrigerator in Exercise 4.13 has a total surface area of $5\,m^2$, how much power is required to maintain the temperature difference between the inside and the outside?

4.15   A string of gondola cars (which may be treated as a single long car) coasts along a level frictionless track. As the cars pass under a loading chute, sand falls vertically into them, at a constant rate. Would you expect the speed of the cars to decrease exponentially with distance traveled? Why?

4.16   When the linear absorption coefficient defined by equation (4.1) is divided by the density of the absorber the result is called the *mass absorption coefficient*, often stated in $cm^2\,g^{-1}$. For low gamma-ray energies this quantity is almost independent of the nature of the absorber and hence also is the mass half-value thickness. Why then are heavy materials such as iron and lead used for shielding of gamma rays and x-rays?

5 Exponential Approach

> *. . . Truth is truth*
> *To th'end of reck'ning.*

William Shakespeare

5.1 Filling a Water Tank

From the situations described in the preceding chapters it is but a short step to the consideration of other situations in which a process leads to an exponential type of increase or decrease, not to infinity or to zero, but rather to some finite level at which the process stabilizes and proceeds no farther thereafter.

For example, let's go back to the case of the water tank already treated in §3.1. We started with the tank filled to the level h_0 and enquired how the level would sink with time after the water was allowed to leave through the outflow pipe. We now propose to take up the complementary problem: that of *filling* the tank, initially empty, from a large reservoir, through the same pipe. The physical arrangement is pictured in figure 5.1. The reservoir is filled to height h_0. We shall assume that its capacity is so large that the filling of the tank does not measurably affect the level in the reservoir. At time $t = 0$ the valve in the pipe is opened and water starts to flow into the tank. As before, we wish to know the height of the water in the tank at any later time.

Qualitatively, what do we expect will happen? As water flows in, the tank will start to fill up. However, as the water level rises, the back pressure at the tank end of the pipe will increase, thus decreasing the pressure difference between its two ends and reducing the flow rate from the reservoir into the tank. Consequently the rise of the water level in the tank, rapid at first, becomes slower and slower, eventually ceasing altogether when, after a long time, the level in the tank becomes equal to the level in the reservoir.

Let us now examine this process in quantitative detail. The rate of change of the volume of water in the tank dV/dt is directly proportional to the difference in pressure between the ends of the inflow pipe. From equation (1.7), $dV/dt = -F(p_2 - p_1)$, where F is the flow conductance of the pipe and $p_2 - p_1$ is the pressure difference in question. In the present situation this pressure difference is $\rho g(h - h_0)$, where ρ is the density of the liquid and g

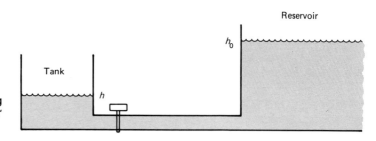

Figure 5.1 The water tank is being filled from a reservoir having 'infinite' capacity.

the acceleration due to gravity. On substitution, $dV/dt = -F\rho g(h - h_0)$. As before, we specify that the tank has vertical sides and crosssectional area A. Then $V = Ah$ and $dV/dt = A\, dh/dt$. Hence

$$\frac{dh}{dt} = \frac{-F\rho g}{A}(h - h_0),$$

and on rearranging,

$$\frac{dh}{h - h_0} = -\frac{F\rho g}{A}dt.$$

Integration gives

$$\ln(h - h_0) = -\frac{F\rho g}{A}t + c. \tag{5.1}$$

To evaluate the constant of integration c, substitute into this equation the initial condition that the tank was empty at the starting time. That is, at $t = 0$, $h = 0$. Hence $c = \ln(-h_0)$, and

$$\ln(h - h_0) = -\frac{F\rho g}{A}t + \ln(-h_0).$$

Taking antilogarithms,

$$h - h_0 = -h_0 e^{-(F\rho g/A)t}$$

or

$$\boxed{h = h_0(1 - e^{-(F\rho g/A)t}).} \tag{5.2}$$

The solid curve of figure 5.2 shows a graph of this result. It confirms that the water level in the tank will rise, rapidly at first and then more and more slowly, approaching eventually the stable level h_0. An asymtotic approach defined by an equation of the form of equation (5.2) is called an *exponential approach*.

As before, the factors in the exponent of equation (5.2) have particular significance. At the instant when $t = A/F\rho g$, the argument of the exponent becomes equal to unity and the water level has risen to the fraction $1 - e^{-1}$ (or $1 - 1/e$) of the eventual equilibrium height h_0. This time is called the one-minus-one-over-e time, the relaxation time, or the *time constant* of the approach function. It is the time required for the approach to go $1 - 1/e$ (or about 63%) of the way to completion.

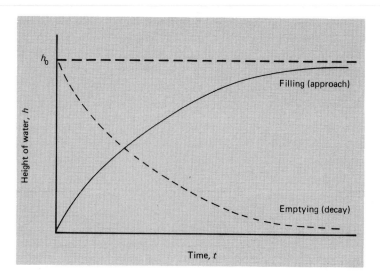

Figure 5.2 The tank fills with an exponential approach to h_0.

▶ **Example 5.1** At what time will the water tank be three-fourths full?

From equation (5.2),

$$h = h_0(1 - e^{-(F\rho g/A)t}).$$

When $h = 0.75 h_0$ we have,

$$0.75 h_0 = h_0(1 - e^{-(F\rho g/A)t}),$$
$$0.25 = e^{-(F\rho g/A)t}.$$

Taking logarithms

$$-1.39 = -(F\rho g/A)t,$$
$$t = 1.39 A/F\rho g.$$

Note the similarity between the exponential approach equation (5.2) and the exponential decay equation for the same water tank emptying through the same pipe from an initial filled level h_0. This decay is plotted as the broken curve in figure 5.2. The time constants are identical, and the shapes of the approach and decay curves are complementary!

To get another perspective on the phenomenon of exponential approach, let us suppose that the tank, instead of being empty at the time the inflow is turned on, is already filled to some initial height h_1. What then? To find out, we return to equation (5.1) and put in the appropriate different boundary conditions to evaluate the constant of integration. Specifically, we now say that at time

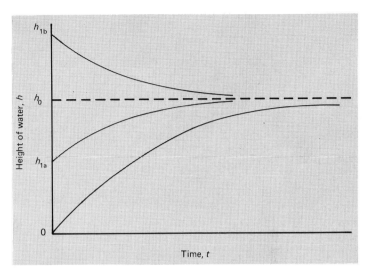

Figure 5.3 The tank always fills up to the level h_0.

$t = 0, h = h_1$. We have on substituting, $\ln (h_1 - h_0) = c$. Hence the solution of the differential equation is

$$\ln (h - h_0) = -\frac{F\rho g}{A}t + \ln (h_1 - h_0),$$

or

$$h - h_0 = (h_1 - h_0)e^{-(F\rho g/A)t}.$$

Solving for h we find:

$$h = h_1 e^{-(F\rho g/A)t} + h_0 (1 - e^{-(F\rho g/A)t}). \qquad (5.3)$$

The first term on the right-hand side of the equation represents the effect of the initial filling level. Note that this effect dies out as time goes on. The second term on the right gives the exponential approach to the final equilibrium level h_0. You see that no matter how much water the tank has in it to start with, it always ends up, after several time constants, with the water level at $h = h_0$. Figure 5.3 shows plots of equation (5.3) for three different initial values of the water level: 0, h_{1a} and h_{1b}.

5.2 Charging a Capacitor

In Chapter 3 we examined the decay of voltage across an electrical capacitor initially charged to some starting voltage V_0 and discharging through a resistor of resistance R. We shall now take up the complementary problem—that of charging the capacitor.

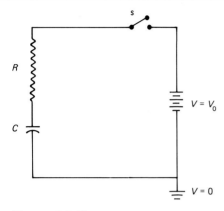

Figure 5.4 The capacitor charges with an exponential approach to V_0.

In the circuit of figure 5.4 we start with the switch s open and the capacitor initially at zero charge and zero voltage. The charging switch is now closed, and charge flows through the resistor and into the upper plate of the capacitor. The voltage on this plate rises, eventually becoming equal to the battery voltage V_0 after a sufficiently long time. We wish to know the voltage across the capacitor at any time t after closing the switch.

We are not going to solve this problem for you. If you are responding at all to the message of this book you will recognize at once that charging the capacitor is the analog of filling the water tank in the last section: the variable voltage V is the analog of the water pressure $\rho g h$ at the bottom of the tank; the constant voltage V_0 of the supply battery is the analog of the steady water pressure $\rho g h_0$ at the bottom of the reservoir; the reciprocal of the resistance R is the analog of the flow conductance F of the pipe. If we tell you, finally, that the capacitance C is the analog of the quantity $A/\rho g$ for the water tank, you should be able to write the solution of this problem by substitution of appropriate electrical quantities into equation (5.2):

$$V = V_0(1 - e^{-t/RC}). \tag{5.4}$$

Instead of filling the tank with water to a certain pressure, we are filling a capacitor with charge to a certain voltage. The exponential approach should give no surprises. The time constant of the charging process, as for the discharging of §3.2, is the product RC.

In case the capacitor is already partly charged to begin with, we can make the approach employed in the last section, leading to equation 5.3, for the water tank already partly filled. By analogy we can state that whatever the initial condition of the capacitor, whether completely uncharged, partly charged, or overcharged to a voltage greater than V_0, the result will always be an exponential approach to V_0.

▶ **Example 5.2** A student measures the capacitance of a capacitor by placing it in a circuit like that of figure 5.4 in series with a 1 MΩ resistor and a 48 V battery. He observes, with an electrostatic voltmeter, that 5 s after closing the switch the voltage across the capacitor is 33 V. What is the capacitance?

From equation (5.4) we have

$$V = V_0(1 - e^{-t/RC}).$$

In this example $V = 33\,\text{V}$, $V_0 = 48\,\text{V}$, $t = 5\,\text{s}$ and $R = 10^6\,\Omega$. Hence,

$$33 = 48(1 - e^{-5/10^6 C}),$$
$$15 = 48\,e^{-5/10^6 C},$$
$$\ln\left(\frac{15}{48}\right) = -5/10^6\,C,$$
$$-1.16 = -5/10^6\,C,$$
$$C = 4.3 \times 10^{-6}\,\text{F}.$$

5.3 Heating a Metal Block

To extend the water-tank analogy to still another case, consider the experimental arrangement shown in figure 5.5. A metal block, initially at temperature $0°C$ is attached to a metal rod having thermal conductance K. The block and the rod are packed in heat-insulating material so that the only way heat can enter or leave the block is by conduction through the length of the rod. At time $t = 0$ the far end of this rod is brought into thermal contact with a metal reservoir containing water boiling at $100°C$. We want to know the temperature of the block at any subsequent time.

Here we have a thermal arrangement which is the analog of the water-tank problem and the charging electrical capacitor problem. Do we need to undertake a lengthy analysis to tell what will happen? We shouldn't. Recognizing the analogy between this problem and its predecessors of the two previous sections, we should be able to write down at once that the temperature of the metal block is going to follow an exponential approach to $100°C$ given by the general expression $T = 100(1 - e^{-t/\tau})$, where τ, the time constant, has yet to be worked out.

To do this, let us look back at the time constant RC for the charging capacitor. Can we find the thermal equivalents for the factors R and C in this expression? We can. If the electrical capacitance of the capacitor is C (in CV^{-1}), then the thermal capacitance Γ of the metal block is the number of joules of heat required to raise its temperature by one degree—its heat capacity. The thermal equivalent of the electrical resistance R is the thermal resistance of the rod, the reciprocal of its thermal conductance K. The time constant τ of the heat flow must then be $\tau = \Gamma/K$, and the thermal approach equation must be

$$T = 100(1 - e^{-(K/\Gamma)t}). \qquad (5.6)$$

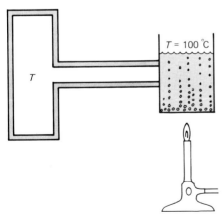

Figure 5.5 The temperature of the block rises with an exponential approach to $100°C$.

5.4 Terminal Velocity

In your introductory physics course you learned that an object dropped from a height will accelerate downward with steadily increasing velocity until it reaches the bottom of its fall. And so it does—in a vacuum. In air, however, or in water, or any other viscous medium, the body will accelerate downward only until the frictional resistance of the medium exerts an opposing force equal to the force of gravity. Thereafter the body falls with a constant velocity called the *terminal velocity*.

Call the mass of the object m and assume that the frictional retarding force is proportional to the velocity v it has acquired. The Newtonian equation of motion then says that the product of the object's mass and its acceleration is equal to the sum of the forces acting on it; that is, the downward force due to gravity mg and the upward frictional retarding force wv, where w is a constant of proportionality dependent on the size and shape of the object and on the viscosity and compressibility of the medium in which it is falling. If we choose the vertical direction as the y direction the equation of motion becomes

$$-m\frac{d^2 y}{dt^2} = -mg + wv, \tag{5.7}$$

where the minus sign denotes the downward sense of the gravitational force and acceleration. Substituting dv/dt for d^2y/dt^2 and dividing through by m, we obtain

$$-\frac{dv}{dt} = -g + \frac{w}{m}v.$$

The solution of this differential equation is

$$v = \frac{m}{w}g + ce^{-(w/m)t},$$

where c is the constant of integration. Assuming that the object falls from rest, the boundary condition is that at $t = 0, v = 0$. On substituting, we obtain

$$v = \frac{m}{w}g(1 - e^{-(w/m)t}). \tag{5.8}$$

This equation says that the object falls with increasing velocity, approaching exponentially the terminal value of $(m/w)g$.

We know that before the frictional retarding force becomes appreciable, the object must begin its fall with the acceleration g.

Let's see if this is consistent with equation (5.8). Differentiating we obtain:

$$a = \frac{dv}{dt} = -\frac{m}{w}g\frac{w}{m}e^{-(w/m)t}.$$

For very small values of t this equation does, indeed, say that the acceleration is equal to g. It also says that at very large values of t, the acceleration is zero; the object has reached terminal velocity.

► **Example 5.3** A locomotive of mass 250 tonnes pulls a train of 50 freight cars, each one with its load having a mass of 100 tonnes. The locomotive exerts a steady pull of 5×10^5 N, and each freight car resists the motion with a drag of $1000 + 500v$ N. What is the terminal velocity of the train? What is the time constant of its approach to terminal velocity?

The equation of motion is

$$\text{mass} \times \text{acceleration} = \text{pull} - \text{drag}.$$

assume 1 tonne = 1000 kg

Therefore,

$$(250 \times 10^3 + 50 \times 100 \times 10^3)\frac{dv}{dt} = 5 \times 10^5 - 50(1000 + 500v),$$

$$5250\frac{dv}{dt} = 450 - 25v.$$

Rearranging and multiplying through by 25 gives

$$\frac{-25dv}{450 - 25v} = \frac{-dt}{210}.$$

Integrating this equation gives

$$\ln(450 - 25v) = \frac{-t}{210} + c,$$

where c is the constant of integration whose value is calculated using the conditions that at $t = 0$, $v = 0$, so $c = \ln 450$. Therefore,

$$\ln(450 - 25v) = \frac{-t}{210} + \ln 450,$$

$$450 - 25v = 450\,e^{-t/210},$$

$$v = 18(1 - e^{-t/210}).$$

From this equation it can be seen that the time constant is 210 s, and for large values of t the terminal velocity is $18\,\text{m}\,\text{s}^{-1}$.

This section is intended to show you that exponential approach is not limited to situations in which a tank is being filled with water, a capacitor is being filled with charge, or a thermal capacitor with heat. You might push that analogy by saying, 'Oh, but the falling object is being filled with velocity!' We assert, however, that the terminal-velocity problem is conceptually distinct from the other three, which are clearly analogs of each other. Further different instances of exponential approach appear in other physical situations. For example, when an AC voltage is suddenly applied to a circuit, it sometimes requires several cycles for the alternating current to come to its final steady value by an exponential approach. A mechanical pendulum activated by certain types of escapement will approach exponentially its final steady state amplitude after being started.

The exponential growth, decay, and approach problems which we have discussed in this and the preceding chapters cover a great deal of physics. We are now ready to leave the subject for the time being, but we shall return later.

▶ **Exercises**

5.1   By analogy with equation (5.3), deduce the expression for the charging of a capacitor which is initially partly charged.

5.2   Show that the time at which the filling and emptying curves cross in figure 5.2 is $0.69 A/F \rho g$.

5.3   An initially empty water tank having unit cross sectional area is receiving a constant input of water at a rate of $I \, \mathrm{m^3 \, s^{-1}}$. At the same time water is running out by gravity from the bottom through a horizontal pipe having flow conductance F. Show that the height of the water in the tank at any subsequent time t is

$$h = \frac{I}{F \rho g}(1 - e^{-F \rho g t}).$$

What is the final water height when the water level has stabilized?

5.4   Suppose the frictional retarding force on a falling object is proportional to the square of its velocity. How would you modify the derivation of §5.4? Would the approach to terminal velocity be exponential?

5.5   Prove by differentiation and substitution that equation (5.8) is a solution of the differential expression of equation (5.7).

5.6   What is the time constant of a 0.001 μF capacitor charging through a 1 MΩ resistance?

5.7   In figure 5.4, suppose an additional resistance R' to be placed in parallel with the capacitor. Show that the charging equation becomes

$$V = \frac{V_0 R'}{R + R'}(1 - e^{-(1/R + 1/R')t}).$$

5.8   A ship's mass is 5×10^6 kg. Its engines and propeller develop a constant thrust of 2×10^6 N. The resistance of the water to the ship's motion is proportional to its velocity v and is equal to $2.5 \times 10^5 v$ $N\,m^{-1}\,s^{-1}$. What is the terminal velocity of the ship through the water? What is the time constant of the approach to the terminal velocity?

5.9   From 1910 to 1972, US coal production remained approximately constant at 5×10^{11} kg/yr. A US Senate-sponsored paper '*US Energy Resources* . . . 1972' gave two estimates of the coal resources remaining in the US: a high figure of 1436×10^{12} and a low figure of 340×10^{12} kg. At what rate would we have to *decrease* consumption of coal below the 1972 rate to make this resource last 'forever'? HINT: Let the rate of consumption be $r = r_0 e^{-(r_0/R)t}$, where R is the total resource.

5.10   Suppose that the water tank of §5.1 is filled from a small reservoir which suffers an appreciable lowering of its own level while filling the tank. How would the derivation of equation (5.2) have to be modified?

5.11   Mention at least two other physical situations, not analyzed in this chapter, where there is an exponential approach to a final end condition.

5.12   An initially pure sample of a radioactive isotope with a very long life is decaying at an essentially constant rate of d disintegrations/s. As it does so, it causes the build-up of a radioactive daughter product which has a very short life with a decay constant of $\alpha\,s^{-1}$. When the daughter has come into equilibrium with the parent it too will decay with d disintegrations/s. Show that the number n of the daughter nuclei present at any time is given by the approach equation,

$$n = \frac{d}{\alpha}(1 - e^{-\alpha t}).$$

5.13   The equation $N = N_0(1 + b)/(1 + be^{-kt})$ has been used to represent the growth of organisms and of populations in an environment providing food at a limited rate. The value of N starts from N_0 and increases to a final value $N_0(1 + b)$. Show that (a) when N is small, growth is approximately exponential, and (b) maximum growth rate occurs when N is half its final value. See §2.5 and figure 2.6. Make a sketch of N against t.

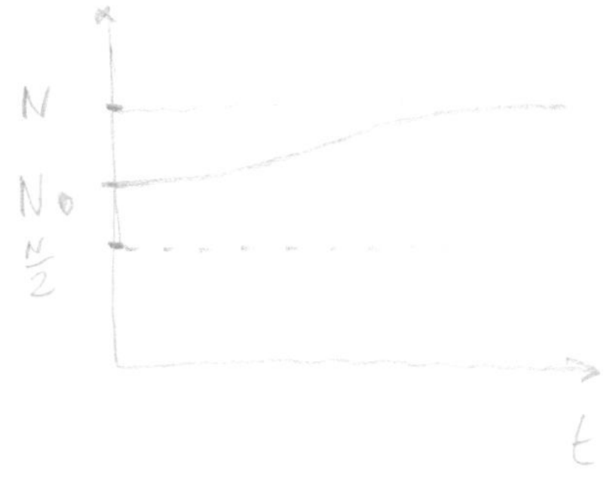

(handwritten margin notes) ? May never happen as illustrated

They mean half way between N_0 and N

6 Oscillations and Oscillators

The mathematician, carried along on his flood of symbols, dealing apparently with purely formal truths, may still reach results of endless importance for our description of the physical universe.

Karl Pearson

6.1 Analogies

We hope that you have inferred from our pursuit of analogies in physics thus far that an analogy is a consistent similarity between equations and structures appearing in two or more fields of knowledge, and that when these similarities are identified they may be used to transfer knowledge of structures and appropriate mathematical procedures of analysis from a familiar field of study to one less well understood.

Analog computers are devices which exploit, for problem solving, analogies between the variables being studied and variables which are conveniently manipulated in the device. The slide rule, invented in the seventeenth century, was probably the first widely-used analog computing aid. Other special-purpose analog computers are the planimeter and various simulators and scale models. The control systems for an oil refinery or the automatic pilot for an aircraft embody many of the features of a general-purpose analog computer.

The general subject of oscillations and oscillators is especially rich in analogies. As we shall point out all oscillators, whether mechanical, electrical, acoustic, or electromagnetic, have certain structural entities in common and exhibit certain features of behavior in common. The purpose of this chapter is to develop these points of similarity so that whenever you contemplate a particular oscillator, for example a pendulum, you will think of it not just as a special kind of mechanical device but also as a member of a much larger and inclusive family that embraces dozens of members varying in size from a vibrating diatomic molecule to an expanding and contracting Cepheid variable star.

6.2 Undamped Oscillations

Consider a simple vibrator consisting of a mass m suspended by a spring from a hook in the ceiling. You pull the mass down for a short distance, then release it. The mass thereupon executes repeatedly an up-and-down motion which continues for a long while. What are the characteristics of this bouncing motion? Well, the *amplitude* of the vibration is clearly at your disposal, depending on the distance you pulled the mass down (or raised it up) before releasing it. However, the *frequency* of the vibration is not yours to choose. The mass exhibits a stubborn determination to vibrate with a particular frequency of its own, called the *natural* frequency f_n.

Let's analyze this situation mathematically. In figure 6.1, let the

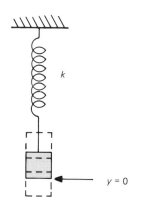

Figure 6.1 The mass oscillates up and down.

rest position of the mass be at $y = 0$, with the vibration occurring in the positive y direction. The Newtonian equation of motion of the mass says that the forces acting upon it must be equal to the mass times the acceleration at any instant of the cycle. In this case the force acting on the mass m is the restoring force of the spring given by $-ky$, where k is the force constant of the spring and y is the vertical displacement of the mass from its rest position. The force constant is defined as the force required to extend or to compress the spring per unit distance. We shall assume that this force constant *is* constant over the amplitude of the vibration. The minus sign indicates that the direction of the restoring force is opposite to the direction of the displacement. Since we have taken $y = 0$ as our reference point, i.e. the position at which the spring's restoring force exactly equals the downward gravitational pull on the mass m, we do not need to include the gravitational force in the following analysis.

The equation of motion is:

$$-ky = m\frac{d^2 y}{dt^2}.$$

The solution of this second-order differential equation is

$$y = c_1 e^{i\sqrt{k/m}t} + c_2 e^{-i\sqrt{k/m}t},$$

where i is the square root of -1, and c_1 and c_2 are constants of integration. The correctness of this solution can be verified by differentiating it twice and substituting into the equation of motion. Expanding this expression we obtain

$$y = (c_1 + c_2)\cos\left[\sqrt{k/m}\,t\right] + (c_1 - c_2)i\sin\left[\sqrt{k/m}\,t\right].$$

If we impose the boundary condition that the mass is released at $t = 0$ when the displacement is maximum and the velocity of the mass is zero, then we obtain the particular solution:

$$y = (c_1 + c_2)\cos\left[\sqrt{k/m}\,t\right].$$

This is the equation of an undamped simple harmonic motion of amplitude $c_1 + c_2$ and natural frequency $f_n = \sqrt{k/m}/2\pi$. If we replace $c_1 + c_2$ by the usual amplitude symbol a we have finally

$$\boxed{y = a\cos\left[\sqrt{k/m}\,t\right].} \qquad (6.1)$$

The natural frequency, you see, is determined by the constants of the system itself; the amplitude a can be whatever you wish to make it. The displacement of the mass is plotted as a function of time in the full line of figure 6.2, starting from $t = 0$ at the instant the mass is released. The velocity of the mass at any instant,

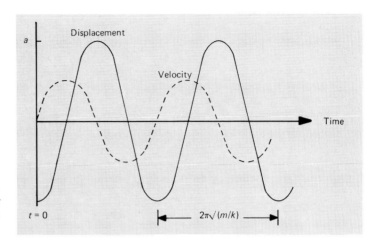

Figure 6.2 The displacement of a mechanical oscillator (full curve) is a quarter of a period behind the velocity (broken curve).

obtained by differentiating the expression for displacement, equation (6.1), is

$$v = a\sqrt{k/m}\,\sin\left[\sqrt{k/m}\,t\right].\qquad(6.2)$$

This velocity is plotted as the broken line in figure 6.2. Note that the displacement is exactly a quarter of a period behind the velocity. When the velocity is maximum the displacement is zero and the mass is going through the center point of its excursion. When the displacement is maximum the velocity is zero and the mass is at the extremity of its motion.

▶ **Example 6.1** A coiled spring requires a force of 1000 N to stretch it 1 cm. What will be the period of vertical oscillation of a 100 kg mass hung from this spring?

The period of vertical oscillation is equal to $1/f_n$, i.e.

$$t = 2\pi\sqrt{(m/k)}.$$

In this example $m = 100\,\text{kg}$ and $k = 1000 \times 100 = 100\,000\,\text{Nm}^{-1}$. Therefore,

$$t = 2\pi\sqrt{(100/100\,000)},$$
$$= 2\pi\sqrt{0.001},$$
$$= 0.20\,\text{s}.$$

All these behavioral features are characteristic of a large class of things called harmonic oscillators. An oscillator is a device or system which is capable of executing a repeated cycle of actions.

FREQUENCIES

Table 6.1 Comparison of some Oscillating Systems

Type of oscillator	How started	Natural frequency, f_n	Compliant element	Inertial element
Spring-and-mass	pull mass down and let go	$\dfrac{1}{2\pi}\sqrt{\dfrac{k}{m}}$	force constant of spring, k	mass
Torsion	rotate disc through an angle and let go	$\dfrac{1}{2\pi}\sqrt{\dfrac{\tau}{k}}$	torque constant of wire, τ, restoring torque of wire per radian twist	moment of inertia, I, of disc
Vibrating string	pluck string sideways and let it go†	$\dfrac{1}{2l}\sqrt{\dfrac{T}{\mu}}$ for fundamental mode	tension T of the string	mass per unit length, μ
L–C electrical circuit	close switch to charge capacitor, then open switch. Charge will oscillate between the plates of the capacitor through inductor L	$\dfrac{1}{2\pi}\sqrt{\dfrac{1/C}{L}}$	compliance, the reciprocal of the capacitance, $1/C$	self inductance L of the inductor

Table 6.1 *contd* . . . .

Type of oscillator	How started	Natural frequency, f_n	Compliant element	Inertial element
Echo chamber	explode firecracker in center of chamber. The chamber will 'ring.'‡	$\dfrac{1}{4l}\sqrt{\dfrac{E}{d}}$	adiabatic volume elasticity, E, of gas in the chamber	density, d, of gas in the chamber
Simple pendulum	draw bob to one side and release it	$\dfrac{1}{2\pi}\sqrt{\dfrac{mg/l}{m}}$ $=\dfrac{1}{2\pi}\sqrt{\dfrac{g}{l}}$	compliance, restoring force per unit displacement of bob, mg/l	mass

Restoring force along arc toward centre of swing $= mg\sin\theta = mg\left(\dfrac{x}{l}\right)$ for small l only

† Plucking the string sideways will cause the string to vibrate at several different frequencies simultaneously. However, the higher frequencies will die out sooner, leaving the string vibrating primarily in its lowest-frequency fundamental mode, as shown.

‡ The firecracker burst will produce a large spectrum of frequencies echoing back and forth in the chamber. However, the walls and ends cannot be made perfectly reflecting, and the higher-frequency components will die out sooner, leaving the sound oscillation in the chamber primarily of the lowest possible frequency, the fundamental mode. This method of starting an oscillation is called 'shock excitation.'

The adjective 'harmonic' means that the time dependence of the action is sinusoidal. The spring-and-mass oscillator we have been describing is clearly such a device, as is a vibrating tuning fork, a capacitor–inductor electrical circuit, a sounding organ pipe, a vibrating diatomic molecule, a freight car rocking back and forth after passing over a bump in the track, and a simple pendulum. There exist also a large number of oscillating motions which are *not* harmonic: the motion of a stick rattling along the pickets of a picket fence, the motion of a bowed violin string emitting a saw-toothed shaped sound wave, the motion of an electron beam scanning the raster of a TV screen, the stop-and-go motion of the escapement of a mechanical clock. These are examples of *anharmonic* oscillations.

In the rest of this chapter we shall consider only harmonic oscillations, since their behavior is easier to analyze mathematically. Also, they embody the basic principles upon which the actions of all kinds of oscillator depend.

We ask now what makes the spring-and-mass oscillator repeat its cycle, once it is set into motion. Starting from the instant when the mass is at the bottom of its excursion, it accelerates towards its rest position because the upward force of the stretched spring exceeds the weight of the mass. As the mass passes through its rest position the upward force of the spring is exactly equal to the weight of the mass, and therefore there is no acceleration at this instant. However, the upward momentum of the mass causes it to coast through the rest position and carries it on towards the highest peak of its excursion. The compressed spring exerts a downward force on the mass, causing it to decelerate and come to rest for an instant before it starts down again to repeat the motion.

The system thus possesses two elements which are required to sustain oscillation: a restoring force which continually urges the mass towards its center rest position whenever it is displaced, and a mass whose momentum causes the overshoot which keeps the oscillation going. Restoring force and mass are found in all oscillating mechanical systems. Their analog equivalents, compliance and inertia, are found in all non-mechanical oscillators. Table 6.1 exemplifies this statement in oscillators of several types: mechanical, electrical, and acoustical.

All these oscillators are similar in their general features. They require some kind of initial action to get them started. They possess some kind of compliance to make the system continually seek its rest position, and some kind of inertia to cause overshooting of the rest position each time it is achieved, twice per cycle. Also, the natural frequency of the oscillation involves in all cases the square root of the ratio of compliance to inertia.

Through analysis similar to that for the spring-and-mass oscillator, it can be shown that in all cases where the restoring force is proportional to the displacement of the system from its rest position *and* is directed towards the rest position, the oscillation will be harmonic.

In all these cases the act of getting the oscillation started endows the oscillating system with an initial amount of energy. This energy at any subsequent time is to be found exchanging back and forth between potential and kinetic forms. In the case of a simple pendulum, for example, the energy is entirely potential every time the mass is at either end of its swing and entirely kinetic each time the mass is going through its center rest position. At all other parts of the cycle the energy is partly potential and partly kinetic, with an ever-changing distribution. The *total* energy of the system, however, is constant throughout the cycle.

6.3 Damped Oscillations

In the previous section we have disregarded the friction which occurs in most oscillating systems and which causes the oscillations to decrease in amplitude and ultimately die out altogether. An actual pendulum encounters friction with the air through which it moves and possible hysteresis in the suspending string. The prongs of a tuning fork radiate sound and this imposes a frictional resistance on the motion. An oscillating electrical circuit loses energy to its own circuit resistance and to whatever radiation is emitted. Almost the only cases of totally undamped oscillation occurring in nature are the vibrations of polyatomic molecules and the vibrations of the atoms of a medium through which radiation is passing without absorption.

We shall now examine in detail the effect of friction on the motion of an oscillator. Let's return to the spring-and-mass system and suppose that the mass is suspended in a dash pot containing a liquid whose density and viscosity present resistance to the motion of the mass. In addition to the term representing the restoring force, the equation of motion must now have a term to represent the retarding force of the friction. Let us suppose that this retarding force is proportional to the velocity of the mass at any instant. The frictional force term will then be

$$-rv = -r\frac{dy}{dt},$$

where r is a resistance constant which depends on the density and viscosity of the liquid, the size and shape of the mass, and the closeness of the side of the vessel. We assume that for the range of velocities involved in the experiment, r is indeed constant. The equation of motion is then

$$-ky - r\frac{dy}{dt} = m\frac{d^2y}{dt^2},$$

and the solution is

$$y = c_1 \exp\left\{\left[-\frac{r}{2m} + \sqrt{\left(\frac{r^2}{4m^2} - \frac{k}{m}\right)}\right]t\right\}$$
$$+ c_2 \exp\left\{\left[-\frac{r}{2m} - \sqrt{\left(\frac{r^2}{4m^2} - \frac{k}{m}\right)}\right]t\right\}. \quad (6.3)$$

As before, this solution can be verified by making the indicated differentiations and substituting them in the equation of motion. Equation (6.3) can lead to any one of three possible outcomes, depending on whether the quantity under the square root sign is positive, zero, or negative. We shall look at these outcomes separately and give the physical interpretation of each.

Case I The liquid in the dash pot is very viscous, r is large, and $r^2/4m^2$ is greater in magnitude than k/m. In this case the quantities under the radical signs will be positive but smaller than $r^2/4m^2$. The radicals themselves will thus be positive and smaller than $r/2m$. Hence the arguments of the exponents will be negative in both terms of equation (6.3), and the solution is simply the algebraic sum of two decay exponentials. After the mass has been pulled down and released, it returns sluggishly to its rest position, restrained by the viscosity of the liquid. It does not overshoot the rest position and therefore it doesn't oscillate at all. The broken line of figure 6.3 shows this behavior, which is said to be *overdamped*.

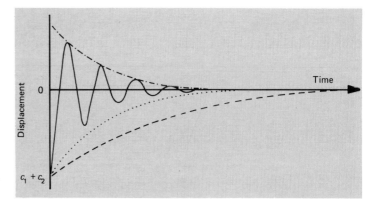

Figure 6.3 The motion of a damped oscillator may be either periodic or non-periodic. Broken curve, overdamped (Case I); dotted curve, critically damped (Case II); full curve, underdamped (Case III); chain curve, decay envelope.

Case II The resistance constant, force constant, and mass are so related that $r^2/4m^2$ is exactly equal to k/m. In this case the quantities under the radical signs in equation (6.3) have zero value and the solution simplifies to

$$y = (c_1 + c_2)\,e^{-(r/2m)t}. \tag{6.4}$$

This solution is also a simple decay exponential, but with a shorter time constant than for the solution of Case I. Again the mass doesn't oscillate, but merely returns exponentially to its rest position after being displaced and released. However, Case II has the quickest return possible without overshooting. The dotted curve of figure 6.3 shows this behavior, which is described as being *critically damped*.

Case III The liquid is of low viscosity and $r^2/4m^2$ is less than k/m. The quantities under the radical signs are negative and therefore the radicals themselves are imaginary. The solution of equation (6.3) then becomes:

$$y = c_1 \exp\left\{\left[-\frac{r}{2m} + i\sqrt{\left(\frac{k}{m} - \frac{r^2}{4m^2}\right)}\right]t\right\}$$
$$+ c_2 \exp\left\{\left[-\frac{r}{2m} - i\sqrt{\left(\frac{k}{m} - \frac{r^2}{4m^2}\right)}\right]t\right\}.$$

By application of Euler's theorem this becomes

$$y = c_1 e^{-(r/2m)t}\left\{\cos\left[\sqrt{\left(\frac{k}{m} - \frac{r^2}{4m^2}\right)}t\right]\right.$$
$$+ i\sin\left[\sqrt{\left(\frac{k}{m} - \frac{r^2}{4m^2}\right)}t\right]\right\}$$
$$+ c_2 e^{-(r/2m)t}\left\{\cos\left[\sqrt{\left(\frac{k}{m} - \frac{r^2}{4m^2}\right)}t\right]\right.$$
$$- i\sin\left[\sqrt{\left(\frac{k}{m} - \frac{r^2}{4m^2}\right)}t\right]\right\},$$
$$= (c_1 + c_2)e^{-(r/2m)t}\cos\left[\sqrt{\left(\frac{k}{m} - \frac{r^2}{4m^2}\right)}t\right]$$
$$+ i(c_1 + c_2)e^{-(r/2m)t}\sin\left[\sqrt{\left(\frac{k}{m} - \frac{r^2}{4m^2}\right)}t\right].$$

As before, let us seek a particular solution satisfying the boundary condition that the mass is released at time $t = 0$ with zero initial velocity from the position of maximum displacement. The situation can then be described by the equation

$$y = (c_1 + c_2)e^{-(r/2m)t}\cos\left[\sqrt{\frac{k}{m} - \frac{r^2}{4m^2}}\,t\right]. \qquad (6.5)$$

In this case the mass does overshoot its rest position when displaced and released. It oscillates with initial amplitude $c_1 + c_2$ and frequency $(1/2\pi)\sqrt{(k/m - r^2/4m^2)}$. Furthermore, the amplitude decays according to the exponential factor $e^{-(r/2m)t}$. The resulting behavior is sketched by the full curve in figure 6.3. This oscillation is described as being *underdamped*. Note that another effect of damping is to decrease the frequency of the oscillator from $(1/2\pi)\sqrt{(k/m)}$ (undamped) to $(1/2\pi)\sqrt{(k/m - r^2/4m^2)}$ (damped). If the damping is removed, the solution of equation (6.5) becomes identical to that of equation (6.1) for the same oscillator without damping, as would be expected.

In many practical applications of oscillators damping is unwanted and design steps are taken to reduce it as much as possible. On the other hand, there are some situations in which damping is deliberately sought. Consider the mounting of the body of an automobile. When oscillations of the car body are initiated by driving over a bump in the road, you do not want the body to continue bouncing up and down for the next quarter of a mile. You want the oscillations to disappear quickly and to restore your smooth ride. Hence shock absorbers (resistance elements attached to the spring-and-mass system of the car) are used to absorb the energy of the oscillation and dampen its amplitude. Some of the more sophisticated shock absorbers are adjustable so that they can be set for critical damping of the particular car in which they are installed.

Another application where damping is deliberately introduced is in instruments of measurement. One does not want to wait indefinitely for the needle of an ammeter or the indicator arm of a balance to settle down to its final reading. Critical damping is provided in most such instruments to get rid of overshoot and permit quick reading.

6.4 Logarithmic Decrement and Quality Factor

A term often used in engineering to describe the rate at which an underdamped oscillation decays is the logarithmic decrement, δ, which is defined as the natural logarithm of the ratio of the amplitude of any oscillation to the amplitude of the one immediately preceding it. This ratio will be given by the magnitude of the decay factor $e^{-(r/2m)t}$ evaluated over one period.

The natural logarithm of this ratio is simply the argument of the exponent so evaluated. Thus the logarithmic decrement is

$$\delta = \frac{r}{2m} \frac{2\pi}{\sqrt{(k/m - r^2/4m^2)}} = \frac{2\pi}{\sqrt{[(4km/r^2) - 1]}}. \qquad (6.6)$$

Another engineering description of the rate of decay of an underdamped oscillation is the quality factor, Q, defined as the number of cycles required for the amplitude of the oscillation to decrease by the factor $e^{-\pi}$. Since with each cycle the amplitude decreases by the factor $e^{-\delta}$ it follows that

$$Q = \pi/\delta = \tfrac{1}{2}\sqrt{[(4km/r^2) - 1]}. \qquad (6.7)$$

In cases where the damping is slight and the first term under the radical sign is very much greater than unity, a little algebraic manipulation shows that

$$Q \simeq \frac{k}{2\pi r f_n}. \qquad (6.8)$$

Another way of conceptualizing the meaning of Q is to think of Q as a measure of the energy stored in the oscillator at any instant in ratio to the energy lost per cycle. Table 6.2 gives values of Q for some typical oscillators. The values are only approximate; they vary widely in each category.

Table 6.2 Some values of Q

Electric L–C circuit	100
Earth, for earthquake waves	250–1400
Tuning fork	4.5×10^4
Plucked violin string	10^3
Cavity microwave resonator	10^4
Quartz crystal	10^6
Excited atom	10^7

6.5 Damped Oscillation with Periodic Driving Force

In some practical applications of oscillators, we require an oscillation which is not damped out but which maintains a constant amplitude. Examples are the carrier signal of a radio broadcasting station, a sustained musical note, the pendulum of a clock, and a beam of light from a lamp. It must be clear that to sustain an oscillation at constant amplitude in the presence of inevitable damping, some mechanism must be provided to supply in each cycle an amount of energy equal to the energy lost per cycle from damping. Thus radio broadcasting antennas are driven by electronic power generators; sustained musical notes are obtained from violin strings by steady bowing rather than by plucking; clock pendulums are driven by spring-operated or weight-operated escapements; the emission of light from a lamp is sustained by the electric power that heats the filament.

We have already shown that an underdamped oscillator, in the absence of any driving mechanism other than the initial impulse, always oscillates at a particular natural frequency f_n determined by its compliant and inertial constants. However, such an oscillator can be forcibly driven at other frequencies as well, by an external driving agency. The resulting motion is called *forced oscillation*.

The response of the oscillator to the driving force depends, among other factors, on the relationship between the frequency of the driver and the natural frequency of the oscillator. To determine this dependence, let us return to the underdamped mass-and-spring oscillator of §6.3. This time, instead of having the spring suspended from a rigid support, let the support itself move up and down with a sinusoidal motion of frequency f_s and constant amplitude. The periodic motion of the support constitutes the driving mechanism. In response, the mass oscillates with two different motions superimposed. First there is motion at the natural frequency of the oscillator—motion which is activated by the initial turning on of the driving mechanism. This natural-frequency oscillation dies out with time at the natural damping rate $e^{-(r/2m)t}$ and shortly disappears. This initial motion is called a *transient*. The second motion is the forced oscillation at the frequency of the driver. The forced oscillation starts when the driving mechanism is turned on and continues so long as the driving force continues. Initially the transient and the forced oscillations are superimposed; the resulting motion is sometimes puzzling to watch. After the transient has damped out and the forced oscillation alone remains, the system settles down to a steady-state condition which we wish now to investigate.

The vertical oscillation of the support introduces a third force term into the equation of motion of the mass. Let this force be represented by the real part of $F_s e^{i2\pi f_s t}$, where F_s is the amplitude of the oscillating force provided by the oscillating support. The differential equation of motion then becomes

$$-ky - r\frac{dy}{dt} + F_s e^{i2\pi f_s t} = m\frac{d^2 y}{dt^2}. \tag{6.9}$$

You may remember that the complete solution of this kind of differential equation is the same as that for the same oscillator *without* the driving force term *plus* a particular integral which, when differentiated and substituted back, satisfies the differential equation. We know from experience that this particular integral, which gives the steady-state motion after the initial transient has died away, must be given by an expression of the form

$$y_s = a_s e^{i2\pi f_s t},$$

having the same frequency as the driving frequency f_s, but in general differing in phase and amplitude from the phase and amplitude of the driver. Substituting this value of y_s into the differential equation, we obtain†

$$-ka_s e^{2\pi i f_s t} - ra_s 2\pi i f_s e^{2\pi i f_s t} + F_s e^{2\pi i f_s t} = -ma_s 4\pi^2 f_s^2 e^{2\pi i f_s t}$$

$$-ka_s - ra_s 2\pi i f_s + F_s = -ma_s 4\pi^2 f_s^2$$

$$a_s = \frac{F_s}{k - 4\pi^2 m f_s^2 + 2\pi i r f_s}$$

$$= \frac{F_s/m}{k/m - 4\pi^2 f_s^2 + 2\pi i r f_s/m}$$

But $k/m = 4\pi^2 f_n^2$, so

$$a_s = \frac{F_s/m}{4\pi^2 \left(f_n^2 - f_s^2 \right) + 2\pi i r f_s/m}. \qquad (6.10)$$

The right-hand side of equation (6.10) is a complex quantity, indicating a complex amplitude for the forced oscillation. The physical interpretation is that the amplitude of the forced motion has a real part in phase with the driver and an imaginary part $90°$ out of phase with the driver. To separate the real and imaginary components of a_s, multiply the numerator and denominator of expression (6.10) by the complex conjugate of the denominator

---

† In our preoccupation with the derivation and final form of the displacement amplitude a_s we should not forget that the *complete* solution of the equation of motion (6.9) is the superposition of the solution (6.5) and the particular integral solution (6.10). That is

$$y = (c_1 + c_2)e^{-(r/2m)t} \cos\left[\sqrt{\frac{k}{m} - \frac{r^2}{4m^2}}\, t \right] + a_s \cos(2\pi f_s t).$$

In what follows, however, we shall be interested only in the steady-state driven motion which continues after the starting transient has died out.

and perform some simplifying algebra, obtaining finally,

$$a_{s,real} = F_s \frac{f_n^2 - f_s^2}{4\pi^2 m (f_n^2 - f_s^2)^2 + r^2 f_s^2 / m},$$

$$a_{s,imag} = -F_s \frac{2\pi r f_s}{[4\pi^2 m (f_n^2 - f_s^2)]^2 + 4\pi^2 r^2 f_s^2}. \tag{6.11}$$

The minus sign of the imaginary component means that the 90°-out-of-phase component of the displacement is lagging behind the phase of the driving force.

Examination of these two expressions reveals the following aspects of the forced oscillation. At very low frequencies, far below the natural frequency f_n of the oscillator, the displacement is almost entirely real and in phase with the driving force. As the driving frequency is increased, both components of the complex amplitude increase. As the driving frequency approaches the natural frequency of the oscillator, the in-phase component makes a steep dip, while the out-of-phase component increases still farther, indicating that the overall phase of the displacement is lagging farther and farther behind the phase of the driving force. As the driving force is made equal to the natural frequency, the in-phase component of the displacement passes through zero, becoming negative, and the displacement is entirely out of phase with the driving force, lagging behind it by 90°. At still higher driving frequencies both components decrease in magnitude, with the in-phase component predominating. However, the in-phase component is now negative in sign, indicating that the phase of the displacement approaches 180° opposition to the driving force at very high frequencies.

These features of the forced oscillations are brought out in the plots of figure 6.4 where $a_{s,real}$ and $a_{s,imag}$ are sketched for comparison. We suggest that you re-read the last paragraph with reference to this figure.

The overall amplitude of the forced vibration, which is what you actually see and measure, is given by the vector resultant of the real and imaginary parts of the complex amplitude: i.e. $a_s = \sqrt{(a_{s,real}^2 + a_{s,imag}^2)}$. This curve is also plotted in figure 6.4. It shows that the steady-state response of an underdamped oscillator to a periodic driving force is maximum when the frequency of the driver is close to the natural frequency of the oscillator. When this condition is fulfilled, the oscillator is in *resonance* with the driver.

Figure 6.4 also includes two broken curves which show the response of the same forced oscillator, as a function of driving

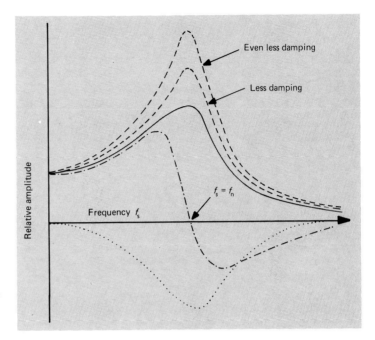

Figure 6.4 The real and imaginary parts of the complex amplitude vary with frequency: full curve, $a_s = \sqrt{(a^2_{s,real} + a^2_{s,imag})}$; chain curve, $a_{s,real}$, dotted curve, $a_{s,imag}$. The responses of the same forced oscillator for reduced amounts of damping are shown by the broken curves.

frequency, for reduced amounts of damping. As the damping is reduced the amplitude of the oscillator at resonance is increased.

6.6 Electrical Oscillating Circuit

The differential expression of equation (6.9) with suitable interpretation of the constants may be regarded as the general equation of motion of *any* kind of underdamped harmonic oscillator with forced motion. Let us see how this generality applies to the case of an electrical circuit (figure 6.5). In the language of electricity, inductance L is the analog of mechanical inertia, resistance R is the equivalent of the mechanical friction element r, and $1/C$ is the compliant element. Thus for an electrical circuit containing inductance, resistance, and capacitance in series with an AC input driving voltage, the equation of motion becomes

$$\frac{Q}{C} - R\frac{dQ}{dt} + V_s e^{2\pi i f_s t} = L\frac{d^2 Q}{dt^2}, \tag{6.12}$$

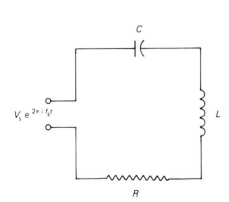

Figure 6.5 This electrical circuit is an analog of the mechanical oscillator.

where Q is the charge on the capacitor at any instant, and the driving voltage is given by the real part of $V_s e^{2\pi i f_s t}$. We could pursue the solution of equation (6.12) just as we did for equation

(6.9), arriving at the real and imaginary components of the complex amplitude of the charge fluctuation on the capacitor. However, electrical engineers are accustomed to thinking in terms of voltage and current, rather than voltage and charge. We can easily recast equation (6.12) to read in voltage and current by differentiating once with respect to t and recognizing that $\mathrm{d}Q/\mathrm{d}t$ is the current I at any instant. We then have

$$-\frac{1}{C}I - R\frac{\mathrm{d}I}{\mathrm{d}t} + 2\pi i f_s V_s e^{2\pi i f_s t} = L\frac{\mathrm{d}^2 I}{\mathrm{d}t^2}.$$

Carrying out the solution of this equation as before, we obtain for the components of the current in phase and out of phase with the driving voltage:

$$I_{real} = \frac{V_s R}{R^2 + (2\pi L f_s - 1/2\pi C f_s)^2},$$

$$I_{imag} = \frac{-V_s(2\pi L f_s - 1/2\pi C f_s)}{R^2 + (2\pi L f_s - 1/2\pi C f_s)^2}. \qquad (6.13)$$

If we define the quantity $2\pi L f_s - 1/2\pi C f_s$ as the *reactance* and designate it by the symbol X we have

$$I_{real} = V_s \frac{R}{R^2 + X^2},$$

$$I_{imag} = -V_s \frac{X}{R^2 + X^2}. \qquad (6.14)$$

The total current, which is what you measure with an AC ammeter, is given by the vector sum of the real and imaginary parts of the current. Thus,

$$I_{total} = V_s \frac{R - iX}{R^2 + X^2} = V_s \frac{R - iX}{(R + iX)(R - iX)} = \frac{V_s}{R + iX}.$$

Note the similarity between this expression and the expression for the current in a DC circuit given by Ohm's law, $I = V_s/R$. In a DC circuit the current is limited by the resistance R, while in an AC circuit it is limited by the quantity $R + iX$. This quantity is called the impedance, Z.

While we have derived it above as the ratio of voltage to current in an electrical circuit, the impedance concept can be extended to include all oscillating systems. In mechanics the impedance would be the ratio of the oscillating driving force to the resulting oscillating velocity of the driven system. In acoustics the impedance of an acoustic element is the ratio of the oscillating pressure to the resulting oscillating velocity of the driven membrane, piston, or medium. Indeed, sound engineers speak of acoustic ohms. In general, impedance can be thought of as the ratio of cause to effect. The concept of impedance is an important one; it will be developed further in Chapter 12.

Figure 6.6 shows plots of the real and imaginary components of the currents for an actual circuit having typical values of L, C, and R. The similarity to the plots of figure 6.4 for the mechanical oscillator is arresting. If you look closely, however, you will see that the real and imaginary roles in the two figures appear to be interchanged. This is because in the electrical case we chose to solve the equation of motion for the current rather than the charge on the capacitor. The current is the electrical analog not of *displacement* but of *velocity*, which is the time derivative of the displacement.

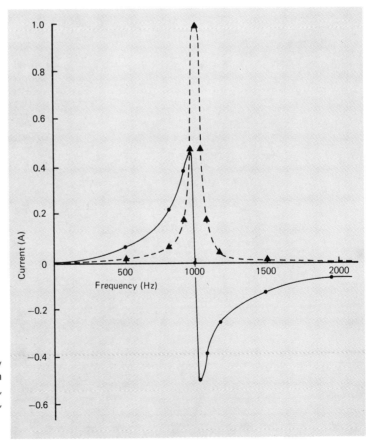

Figure 6.6 The real and imaginary components of the current vary with frequency: full curve, I_{imag}; broken curve, I_{real}. $V_s = 100\,\text{V}$, $R = 100\,\Omega$, $2\pi C = 1\,\mu\text{F}$, $2\pi L = 1\,\text{H}$, $f_n = 1000\,\text{Hz}$.

► **Example 6.2** At what driving frequency does the electric circuit of figure 6.5 become resonant?

The circuit is resonant (exhibits maximum current amplitude) when the denominator of the real and imaginary expressions of equation (6.13) is a minimum, that is, when

$$2\pi L f_s - \frac{1}{2\pi C f_s} = 0,$$

or

$$f_s = \frac{1}{2\pi} \sqrt{\frac{1}{LC}}.$$

Observe that at this resonant frequency the numerator of the imaginary part of the current amplitude vanishes. The current at resonance is therefore entirely real and in phase with the input voltage.

The curves of figure 6.6 show the features of physical behavior for the electrical circuit corresponding to those we derived earlier for the mechanical oscillator. At zero frequency there is no current, in-phase or out-of-phase. As the frequency increases from zero both components increase, with the out-of-phase component predominating and having positive sign, indicating that the current is leading the voltage. As the natural frequency is approached, the out-of-phase component peaks in magnitude, then drops through zero at resonance and becomes negative at still higher frequencies.

The current resonates at a driving frequency equal to the natural frequency of the circuit. At resonance the current is entirely in phase, the out-of-phase component is zero. At resonance the power being absorbed from the AC driving source is maximum, since only in-phase current can deliver power. Finally, the current becomes smaller and smaller as the driving frequency is increased beyond the resonant frequency. In this frequency range the out-of-phase component again predominates in magnitude and is negative in sign, indicating that the current lags behind the voltage, with the lag approaching 90° at very high frequencies.

Additional insight may be had into the workings of this circuit by realizing that at frequencies well below the resonant frequency the current is limited primarily by the reactance of the capacitor. A capacitor will conduct no current at all at zero frequency and only small, out-of-phase current at low frequencies. For frequencies well above the resonant frequency the current is limited by the reactance of the inductor, which increases with increasing frequency. At resonance the capacitive and inductive reactances offset each other, giving a total circuit reactance of zero. In this case the current is entirely real and is limited by resistance alone. At this one frequency the current is given by $I = V_s/R$, as it would be for a simple DC circuit without the capacitor or the inductor.

Exploitation of these properties is nowhere better displayed than in the tuning of a common type of radio receiver. Very weak electrical voltages, at the carrier frequency of the broadcasting station, are picked up by the antenna. These voltages constitute the driving force for the tuning circuit, which consists of a capacitor and inductor in series. When the tuning circuit frequency f_n is adjusted to match the carrier frequency of the particular station being tuned in, the circuit resonates to that carrier frequency; the current in the inductor becomes abnormally large. Other carrier frequencies also picked up by the antenna produce only feeble currents in the tuned circuit because the circuit is not resonant to these frequencies. The tuning circuit thus exhibits selectivity, enabling it to pick out and enhance a single frequency from the many incoming radio frequencies and to ignore all others. The signal thus enhanced by resonance is then passed on to the rest of the receiver for further processing.

In performing this selective function the tuning circuit behaves as a band-pass frequency filter, which we shall discuss further in Chapter 11.

▶ **Example 6.3** An amateur radio operator is instructed by the licensing board to use a carrier frequency of 20.55 MHz. He builds his transmitter with a tuning circuit like that of figure 6.5 with an inductance of 20.00 µH and negligible resistance. What capacitance must he use?

From Example 6.2 we have

$$f = \frac{1}{2\pi}\sqrt{\frac{1}{LC}},$$

$$\sqrt{LC} = \frac{1}{2\pi f},$$

$$C = \frac{1}{4\pi^2 f^2 L}.$$

In this example $f = 20.55 \times 10^6$ Hz and $L = 20 \times 10^{-6}$ H, therefore:

$$C = \frac{1}{4\pi^2 (20.55 \times 10^6)^2 (20 \times 10^{-6})},$$

$$= 3.00 \times 10^{-12}.$$

Therefore a capacitance of 3.00 pF should be used.

6.7 Self Tuning

In many engineering applications of sustained oscillations, provision is made for the frequency of the driver to be controled by the natural frequency of the oscillator. The oscillator thus runs always at its own resonant frequency. The control is exercised through *feedback*, a process whereby some of the output of the oscillator is fed back to control the timing of the input driving force that keeps the oscillator going. The balance wheel and escapement of a mechanical alarm clock furnish an excellent example of a feedback oscillator. The motion of the balance wheel itself determines the instants at which fresh impulses are fed into the wheel to keep it oscillating at its own natural frequency. Similarly, an electric doorbell buzzer, a child's swing, a bowed violin string, a sounding organ pipe, the electronic oscillator of a broadcasting station are other examples of systems which control the frequencies of their drivers. Such systems are called self-tuning oscillators, and are usually employed when some timing function is needed, for example to run a clock, to sustain a musical note, or to produce a constant carrier frequency for a radio station. Note that in most of the cases mentioned above, the oscillator itself is harmonic, while the driving energy is furnished in periodic, but non-harmonic, pulses.

6.8 Oscillators Everywhere

We hope that you have learned much from the material of this chapter about the ways of oscillators, but we hope, even more, that you have seen beyond the specific details of the particular mechanical and electrical examples we have chosen to treat in detail and have, in your mind's eye, scanned the array of oscillators of all kinds, from water sloshing back and forth in the bath tub to the vibrating wing covers of a chirping cricket. All these oscillators have similarities, whether they be mechanical, electrical, acoustical, or optical. They all exhibit the common components of inertia and compliance and the common behavioral properties of natural frequency and resonance. Most of them also exhibit damping. Some are continuously driven rather than initially being started and allowed to vibrate freely. Size has nothing to do with these similarities. We have as tiny oscillators the individual ions of a crystal lattice through which light is shining. The rapidly alternating electric field of the light causes the ions to execute forced vibrations. Such oscillations are of the order of 10^{-8} cm in amplitude. On the other hand we have oscillators of tremendous size, for example the Cepheid variable stars which are expanding and contracting by the inertia of their outer mass and the compliance of their internal radiation pressure. Indeed, according to one interpretation of the 'big bang' theory, the entire universe may be one enormous oscillator with a period of several tens of thousands of millions of years! Wherever you find them and whatever their nature, oscillators are all part of the same brotherhood.

▶ **Exercises**

6.1   Why do motions which are approximately simple harmonic occur frequently in nature? Give some examples. Why are true simple harmonic motions rare?

6.2   Under what conditions will the addition of two simple harmonic oscillations give a resultant which is a simple harmonic oscillation?

6.3   In the derivation of equation (6.1) for the motion of a spring-and-mass oscillator, the mass of the spring itself was disregarded. How would the solution be affected if the mass of the spring were an appreciable fraction of the suspended mass?

6.4   Estimate the Q factor of (a) a playground swing, (b) a simple string pendulum 1.0 m long, (c) the rocking of an empty rocking chair.

6.5   In figure 6.1 the equilibrium $y = 0$ was taken as the position of mass m when the spring was extended just enough so that its restoring force balanced the gravitational force mg. Then the entire potential energy of the system was assumed to be due solely to stretching the spring from this equilibrium position. Suppose now that the equilibrium $y' = 0$ is taken as the lowest point of the *unstretched* spring (no mass attached). The system represented in figure 6.1 will then have gravitational potential energy, to which more potential energy is added as the spring is stretched farther. Show that the motion described for equilibrium at $y' = 0$ is simple harmonic.

6.6   If for the simple harmonic motion of a particular system one doubles the amplitude, by what factor does one change (a) the total mechanical energy, (b) the maximum speed, and (c) the period?

6.7   To minimize the roll of a ship, it may be fitted with an interior tank partly filled with water, (see figure 6.7). The dimensions of the tank are chosen so that the frequency of water sloshing in it matches the roll frequency of the ship. Why does the presence of this tank lessen rather than augment the amplitude of the ship's roll?

6.8   For an exhibit on the theme of *time*, a man constructed working models of various devices for measuring time. To make the size

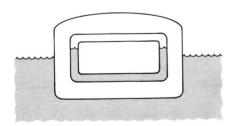

Figure 6.7

impressive he copied existing timepieces scaled up four times in all dimensions. A few hours before the opening of the exhibit all timepieces were set to the correct time and started, but when the curtains were opened, the exhibitor was embarrassed to see discrepancies in the readings. For each instrument state what effect you think scaling up the size had on its rate:

pendulum clock,
torsion pendulum clock,
watch (balance wheel),
watch (tuning fork),
burning candle, marked in hours,
watch (quartz crystal with frequency divider),
hour glass (sand),
sundial,
waterclock.

6.9   If in the arrangement of figure 6.1, the spring exerts a force given by $F = -ay - by^3$, would you expect the period of the oscillation to increase or to decrease with increasing amplitude?

6.10   If a pocket watch is suspended freely by a thread attached to its stem, the watch may begin to swing and its time-keeping rate may change, perhaps by as much as 10 min/day. Why does this effect occur? As a result, why do some watches run faster, others slower?

6.11   A rod of cross sectional area A, length L and average density ρ floats *upright* in a liquid of density ρ_1. Calculate the frequency of the rod bobbing in the vertical direction.

6.12   What is the quality factor for the oscillations with small damping represented by the equation

$$L\frac{d^2Q}{dt^2} + R\frac{dQ}{dt} + \frac{1}{C}Q = 0,$$

where Q is the charge on the capacitor?

6.13   For forced oscillations of an oscillator with no damping, the second term of equation (6.9) vanishes and the equation may be written as

$$\frac{m}{k}\frac{d^2y}{dt^2} + y = \frac{F_0}{k}e^{i2\pi f_0 t},$$

$$\frac{d^2y}{dt^2} + \omega^2 y = \frac{F_0}{m}\sin \omega' t,$$

where $\omega^2 = k/m$, F_0 is the maximum magnitude of the force, and ω' is the angular frequency of the external force. Show that the average power input to the oscillator is zero unless $\omega' = \omega$, the natural frequency.

6.14   In a certain oscillatory motion the amplitude decreases with time; it is 10.0 cm at $t = 0$ and 0.10 cm at $t = 100$ s. A plot of amplitude against time on semilogarithmic paper gives a straight line. (a) Find the equation relating amplitude a and time t. (b) The period is measured as

15.71 s. Calculate the period that this system would have in the absence of damping.

6.15   For displacements on one side of equilibrium a certain spring exerts a force $F = -k_1 x$, but for displacements on the other side $F = -k_2 x$. If this spring is attached to a mass m (figure 6.1) and set into oscillation with small amplitude, show that the period t is $\pi(\sqrt{k_1} + \sqrt{k_2})/m$.

6.16   For an *RCL* series circuit there are two values of frequency, one above resonance f_0 and the other below, for which the power is half the peak value and I^2 is half the resonance value. The difference between these frequencies Δf is called the *bandwidth* of the circuit. The *quality factor Q* was originally *defined* as $Q = 2\pi f_0 L/R$. Show that also $Q = 2\pi f_0/\Delta f$, a useful relation applicable to any kind of resonator.

6.17   Consider damped simple harmonic motion in which the retarding force due to friction or air resistance is proportional to velocity, i.e. the total force $F = -kx - \gamma v$. Show that $x = a\,e^{-\alpha t}\sin(\omega_1 t + \theta)$ is a solution for the differential equation of motion provided that $\alpha = \gamma/2m$ and $\omega = \sqrt{(k - \gamma^2/4m^2)}$, where $\omega = 2\pi f$ is the angular frequency.

6.18   In a certain diatomic molecule the force of interaction between the atoms is given by $F = -a/r^2 + b/r^3$, where r is the separation of the atoms, and a and b are positive constants. (a) What is the equilibrium separation? (b) What is the effective spring constant for small displacements from equilibrium? (c) What is the period?

7 Waves

The reasonable man adapts himself to the world; the unreasonable one persists in trying to adapt the world to himself. Therefore all progress depends on the unreasonable man.

George Bernard Shaw
Man and Superman, III

7.1 The Wave Equation

The nice thing about waves is that they all behave essentially in the same way, whether they be waves of transverse displacement traveling along a stretched string, waves of electric voltage traveling along a transmission line, waves of longitudinal displacement (sound waves) traveling through the air, compressional or shear waves traveling through the rocks of the Earth, torsional waves traveling along a metal bar, waves of electric or magnetic field strength traveling through empty space, or water waves traveling along the surface of the ocean. It will be the purpose of this chapter to explore the behavior of waves in these different physical systems and to show in which aspects they act in similar fashion, whatever their nature.

We shall assume that you are sufficiently familiar with the subject to know what a wave is, and that you understand the relationships between period, frequency, and wave velocity for periodic waves. You are ready for headier stuff. Let's begin by considering a partial differential equation which describes waves of all kinds:

$$\frac{\partial^2 P}{\partial x^2} = \frac{1}{v^2}\frac{\partial^2 P}{\partial t^2}. \tag{7.1}$$

In this equation x is distance along the direction in which the waves are traveling, $1/v^2$ is a constant of proportionality, t is time, and P is a quantity having a particular meaning appropriate to the kind of wave we are talking about. For example, when this equation is used to describe a transverse wave traveling along a stretched string, P is the varying instantaneous sideways displacement of a particle in the string as the wave travels past. If the equation is applied to describe a sound wave traveling through the air, P is the varying longitudinal displacement of a layer of air as the sound wave travels through it. If we are talking about an electromagnetic wave, P is the instantaneous strength of the transverse electric or magnetic field at some point in space through which the wave is passing.

What does this equation say? Let us select as an example the case of a transverse wave traveling along a string. The term $\partial^2 P/\partial x^2$ is the *curvature* of the wave *shape* at any point. The term $\partial^2 P/\partial t^2$ gives the transverse *acceleration* of a particle in the string at that point. The wave equation, then, is a mathematical statement that the curvature of the wave shape at any place and instant is proportional to the sideways acceleration of the string at that place and instant. We shall show presently that the constant of proportionality $1/v^2$ is the reciprocal of the square of the wave velocity along the string. Equation (7.1) is sufficiently general to be applicable to a wave of any sort. Whenever, in your reading, you come across a second-order partial differential equation that looks like equation (7.1), you should be able to say 'Ha, I smell a wave!' It is evident from what we have just said that the quantity P depends both on location x and on time t. Therefore, the wave described by equation (7.1) varies both with location and with time. That is, it has a *shape* and a *motion*.

It is not our intention here to derive equation (7.1). Its derivation, for the physical systems in which waves are found, depends on the kind of wave we're talking about and on what it is that's oscillating. These derivations can be found in other textbooks on waves. Our purpose here is to develop the commonality of waves of all kinds. Nor shall we labor the mechanics of solving equation (7.1) for various situations. Rather, we shall simply state that the general solution is:

$$P = f_1(x - vt) + f_2(x + vt). \tag{7.2}$$

The first term on the right-hand side represents a wave traveling along the positive x direction; the second term represents a wave traveling along the negative x direction.† These two waves may have any shape, as defined by the forms of the functions f_1 and f_2. In some cases the wave or waves may be traveling in one direction only, so that only one of the two terms on the right-hand side of equation (7.2) will be appropriate. Also, the disturbance may be a single wave or pulse, or it may be repeated, as when we have a continuous train of periodic waves.

To pin these generalities down, let's consider the particular case of a wave traveling in one direction only, from left to right in the positive x direction, along an infinitely-long stretched string. Equation (7.2) then becomes $P = f_1(x - vt)$. In this case the quantity P is the instantaneous sideways displacement from its normal position of a short length element dl of the string. If we take a snapshot of this wave at time $t = 0$, the plot of P against x

---

† We here interject a comment on the sign of the quantity vt in equation (7.1). If we are contemplating a point of constant phase (e.g., the peak) on a wave as it moves along, it is clear that P must remain constant with time to preserve the wave shape. If the wave is moving toward the right, x is increasing positively. The product vt must therefore have a *negative* sign to preserve the constancy of the quantity $(x - vt)$. If $(x - vt)$ is to remain constant, then by differentiation $dx/dt - v = 0$, and v is seen to be the velocity of a point of constant phase. It is called the *phase velocity*.

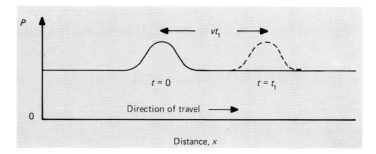

Figure 7.1 The wave has both shape and motion.

may have the appearance of the full curve in figure 7.1. This curve is given by $f_1(x)$, which defines its shape. If we take another snapshot of the same wave at a later time $t = t_1$, we find that it has moved to the right, by a distance vt_1, without changing its shape.

Most of the wave situations having engineering interest involve continuous periodic waves. The simplest kind of periodic wave to conceptualize, as well as to deal with mathematically, is a wave having the shape of a sine curve. For this sinusoidal wave the shape function f_1 is the sine function, and

$$P = a \sin\left(\frac{2\pi}{\lambda}(x - vt)\right), \tag{7.3}$$

where a is an arbitrary constant specifying the amplitude of the wave. The factor $2\pi/\lambda$ has been introduced to keep the argument of the sine function dimensionless, λ itself having the dimension of length. An instantaneous snapshot of this wave at time $t = 0$ is given by the full curve in figure 7.2. The equation of this curve is

$$P = a \sin\left(\frac{2\pi x}{\lambda}\right).$$

If, with t still at zero, we go in the positive x direction from $x = 0$ to $x = \lambda$, the sine function goes through one complete cycle. Thus

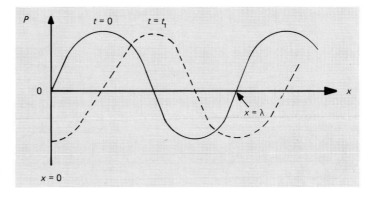

Figure 7.2 The wave moves to the right with velocity v.

λ is seen to be the repeating distance, or wavelength, of the wave.

Another instantaneous snapshot of the same wave at some later time t_1 would show a curve given by:

$$P = a \sin\left(\frac{2\pi}{\lambda}(x - vt_1)\right),$$

$$= a\left[\sin\left(\frac{2\pi x}{\lambda}\right)\cos\left(\frac{2\pi vt_1}{\lambda}\right) - \cos\left(\frac{2\pi x}{\lambda}\right)\sin\left(\frac{2\pi vt_1}{\lambda}\right)\right].$$

If in this expression we let t_1 be a quarter of a period, for example, for which $t_1 = \lambda/4v$, this equation simplifies to $P = -\cos(2\pi x/\lambda)$, as given by the broken curve of figure 7.2. The broken curve is simply a replica of the full curve advanced by a quarter of a wavelength in distance and also by a quarter of a period in time.

To gain further insight into the nature of this wave, let us look not only at the shape of the wave, as we have just done, but also at the time variation of the displacement P at some fixed value of x in space. For example, taking $x = 0$, equation (7.3) becomes $P = -a \sin(2\pi vt/\lambda)$. The displacement is thus shown to have a sinusoidal oscillation with time as well as a sinusoidal shape along the string. The oscillation has frequency given by v/λ.

7.2 Velocity of Propagation

Note that the wave equation (7.1) includes a quantity v which we have shown to be the velocity of propagation of the wave. Our intuition tells us that this wave velocity should depend on the nature of the wave (whether it is mechanical, electrical, etc) and on the physical properties of the medium in which it is traveling. This is true. In the derivation of the wave equations for various wave systems in their various natural media, the velocity does indeed depend on the properties of the medium. For instance, the velocity of a sound wave traveling in a compressible gas can be shown to be

$$v = \sqrt{\frac{E}{d}}, \tag{7.4}$$

where E is the adiabatic volume elasticity and d the density of the gas. The adiabatic volume elasticity is defined as the ratio $-V \, dp/dV$, where p is the pressure and V the volume, and the measurements determining its value are taken in such a way that heat cannot enter or leave the test volume as it is compressed or relaxed.† The minus sign is included to give a positive sign to the elasticity, since according to Boyle's law dp/dV is itself negative in sign.

The velocity of a sound wave is thus seen to be given by the square root of the ratio of the elasticity to the density of the transmission medium. Do waves of other kinds exhibit analogous relationships for wave velocity? Let's see. What about a longitudinal compressional wave traveling along a spring? For this kind of wave the velocity can be shown to be

$$v = \sqrt{\frac{\kappa}{\mu}}, \tag{7.5}$$

where the linear elasticity κ is given by the ratio $l\,\mathrm{d}F/\mathrm{d}l$, where l is the unstretched length of the spring and F is any force compressing or stretching it. The linear density of the spring, μ, is its mass per unit length. If we enquire about the velocity of a longitudinal wave traveling along a metal rod from one end to the other, we have

$$v = \sqrt{\frac{Y}{\mu}}, \tag{7.6}$$

where Y is the Young's modulus and μ the mass of the rod per unit length. For a current or voltage wave traveling along a two-wire transmission line, the wave velocity is

$$v = \sqrt{\frac{1/C}{L}}, \tag{7.7}$$

where C is the capacitance between the wires per unit distance along the line and L is the inductance per unit distance. In comparing electrical and mechanical systems, we recall that the electrical quantity analogous to compliance is capacitance, and the quantity analogous to mass is inductance.

If we are dealing with a torsional wave of twisting and untwisting traveling along a cylindrical rod, we can show that the wave velocity is

$$v = \sqrt{\frac{\tau}{I}}, \tag{7.8}$$

where τ is the torsion constant of the rod, (that is, the torque required to produce one radian of twist per unit length of the rod), and I is the rotational moment of inertia per unit length. For

† This elasticity may be recognized as the reciprocal of the compressibility, defined as the fractional increase in volume per unit increase in pressure.

a transverse vibrational wave traveling along a stretched string the velocity is

$$v = \sqrt{\frac{T}{\mu}}, \tag{7.9}$$

where T is the tension of the string, which is a measure of its transverse compliance, and μ is the mass per unit length along the string.

When we collect all these velocity expressions together and examine them for similarities, we find the following striking collection:

$$v = \begin{cases} \sqrt{(E/d)} & \text{sound waves in a gas,} \\ \sqrt{(\kappa/\mu)} & \text{compressional waves on a spring,} \\ \sqrt{(Y/\mu)} & \text{longitudinal waves in a metal rod,} \\ \sqrt{\dfrac{1/C}{L}} & \text{electric waves on a transmission line,} \\ \sqrt{(\tau/I)} & \text{torsional waves on a metal rod,} \\ \sqrt{(T/\mu)} & \text{transverse waves on a string.} \end{cases} \tag{7.10}$$

For the set of equations above, the symbols differ from wave system to wave system, but behold, in all cases the wave velocity is given by the square root of the ratio of a quantity defining the springiness of the medium to a quantity defining its inertia! Nature is here showing you a welcome consistency.

▶ **Example 7.1**   A flexible steel wire of diameter 0.80 mm and density 7.8 g cm^{-3} supports a 3.6 kg load. What is the speed of a transverse wave in this wire?

From equation (7.9),

$$v = \sqrt{\frac{T}{\mu}}$$

$$= \sqrt{\frac{mg}{\rho \pi r^2}}.$$

In this example $m = 3600$ g, $\rho = 7.8 \times 10^6$ g m^{-3}, $r = 0.4 \times 10^{-3}$ m, and $g = 9.81$ m s^{-2}. Therefore

$$v = \sqrt{\frac{3600 \times 9.81}{7.8 \times 10^6 \times \pi \times (0.4 \times 10^{-3})^2}},$$

$$= 95 \text{ m s}^{-1}.$$

7.3 Energy Transport by Waves

Another property which waves of all kinds exhibit in common is that they are carriers of energy. It requires energy to produce a wave in an initially uniform medium. This energy travels along with the wave, and is either dissipated as heat along the way or else delivered in some form somewhere else, where it may be absorbed to do something useful or may be reflected back towards the source.

Let us review the operation of these several factors in the case of an AC generator sending waves of current and voltage along a two-wire transmission line. A generator can be looked upon, somewhat naïvely, as an electron pump. Suppose we consider the situation at the instant the generator is turned on and has just begun to pump a surge of electrons into the left-hand end of the upper wire (figure 7.3), giving a negative charge to this region of the wire. The electrons required for this process are obtained by pumping them out of the left-hand end of the lower wire, leaving it positively charged. It requires an expenditure of energy to perform this separation of positive charge from negative charge, energy which must be furnished by whatever is driving the generator. The energy thereafter resides in the electric field between the wires.

Electric charge has a tendency to move from regions which are strongly charged to adjacent regions which are neutral. The positive and negative charges which are pumped into the wires by the generator therefore move towards the right, transferring the condition of being charged from point to point along the transmission line and carrying the energy of the field along with them.

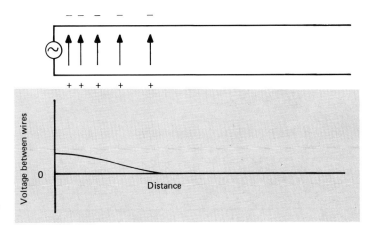

Figure 7.3 The generator is transferring electrons from the lower wire to the upper wire.

Half a cycle later, the polarity of the generator has reversed and electrons are drawn *out* of the left-hand end of the upper wire and pumped *into* the input end of the lower wire (figure 7.4). The electric field between the two wires immediately adjacent to the generator therefore reverses direction. This new situation then propagates to the right along the wires, following half a wavelength behind the situation previously described. With each succeeding cycle of the generator the entire sequence of processes just described is repeated; the resulting waves move along the transmission line with the velocity $v = \sqrt{(1/LC)}$.

It must not be thought that the *electrons* move through the wires with the same velocity as the waves. Wave velocities on typical transmission lines are about a tenth of the velocity of light in free space; the electrons themselves may move with velocities of the order of only a few centimetres per second. The moving electrons transmit their motion by electrostatic repulsion to the electrons in the adjacent length of wire, just as during the passage of a sound wave the moving molecules in the adjoining layers of gas transmit pressure from layer to layer. In both cases the velocity of transmission far exceeds the velocity of the moving particles themselves. Similarly, in a long freight train starting from rest, the cars may be initially jerked from standstill to a velocity of $1 \ km \ h^{-1}$, while the wave of taking up the slack in the couplings may travel backwards from the locomotive, as a wave of rarefaction, with a velocity of $200 \ km \ h^{-1}$! In the wires of the transmission line the electrons merely oscillate back and forth over short distances along the wires as the waves pass over them.

As the electrons jostle back and forth in the wires they constitute an alternating current. As charge flows through the resistance of the wires, heat is generated, and the energy

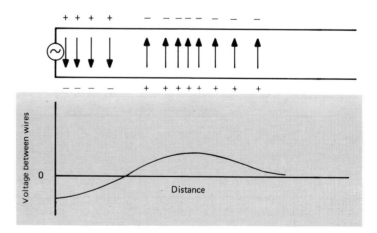

Figure 7.4 This situation follows that of figure 7.3 by half a period.

represented by this heat must be abstracted from the energy of the waves, leading to the exponential decay of the current and voltage as described in detail in Chapter 4 for a DC transmission line. The argument developed there loses none of its validity when applied to AC transmission.

At the other end of the line the wave energy may be delivered to some load—an electrical device, lamp, motor, or appliance, performing a useful function. In general this output load will absorb only a portion of the available wave energy coming along the line. In such a case the unused portion of the energy will be reflected back toward the generator in the form of waves traveling in the opposite direction and having the same frequency, velocity, and wave shape as the original waves, but with diminished amplitude appropriate to the fraction of the original energy absorbed by the terminal load. These reflected waves are described by the second term on the right-hand side of equation (7.2), until now disregarded. We shall resume discussion of the phenomenon of wave reflection in §7.4.

To drive home to you the universality of the wave-energy ideas presented above for the specific case of an electrical transmission line, we shall show how the story goes for the analogous situation of a sound wave propagating along the inside of a hollow tube, which thus becomes an acoustic transmission line. Imagine the sound source to be a piston in the input end of the tube actuated by a mechanical driver of some sort (figure 7.5). The sound waves are carriers of the energy which resides in the compressions and rarefactions of the waves and travels along with them. This energy was impressed upon the waves by the motion of the piston, which, in turn, derived the energy from the work done by the driving mechanism. In detail, here's how it goes. Consider an instant when the piston is moving towards the right and piling up a compression in the layers of air immediately ahead of it. At the same instant the rear surface of the piston is creating a partial vacuum—a rarefaction. The piston does work on the air by moving to the right against the net force opposing its motion.

The compression, once created, moves off down the tube to the right, each successive layer of air in the tube passing the compression on to the next adjacent layer to the right. The energy

Figure 7.5 The oscillating piston sends waves of compression, and rarefaction along the tube.

of compression travels along with the compression. Half a cycle later the piston is moving to the *left*, creating a region of rarefaction in the input end of the tube and doing more work in the process. The rarefaction then follows the previous compression down the tube to the right. The entire sequence is repeated cycle after cycle.

During the transmission of the sound waves down the tube, some of the energy is lost in the following ways. (i) The layers of air which at any instant are under compression are at a slightly higher temperature than the adjacent regions of rarefaction. Heat is therefore continually leaking out of the compressions into the rarefactions. When heat is removed from a compressed layer of gas, its pressure decreases, and the heat flowing into a layer of rarefied gas warms it slightly and increases its pressure. The heat leakage therefore progressively reduces the amplitude of the pressure fluctuations of the wave as it travels along, with consequent reduction of the energy it carries. (ii) The sidewalls of the tube absorb some heat from the passing compressions and deliver some of the absorbed heat to the rarefactions, with the same result as above. (iii) The increased temperature in the compressions may cause an increase in the internal rotational and vibrational energy of the gas molecules. Some of this extra energy may be retained for half a cycle and delivered as heat during the succeeding rarefaction, thus warming it and reducing its amplitude. The processes of compression and rarefaction are not strictly adiabatic.

These processes have the same final result as the processes whereby electrical energy is dissipated when waves of voltage and current travel along a transmission line. Indeed, you may think of them as constituting a kind of acoustical resistance.

At the far end of the tube the wave energy may be partly absorbed by a listening ear. On the other hand, if the output end of the tube is completely unobstructed, part of the sound energy will pass out into the surrounding air and the rest will be reflected back along the tube towards the generator. If the output end is completely blocked by a rigid non-absorbing baffle, the sound energy will be totally reflected back along the tube.

We hope you will think some more about the parallel features presented by the electrical and acoustical wave systems we have described in this section since they are typical of all wave systems. The common features are these: (i) energy is required to produce a wave; (ii) the energy used to produce it travels along with the wave; (iii) this energy is progressively lost by some kind of frictional mechanism as the wave travels along; and (iv) the energy that remains may be partly or totally absorbed by some energy-

converting process at the end of the transmission path, or it may
be partly or totally reflected. Since energy can be neither created
nor destroyed, all of the energy imparted to the wave in the act of
producing it must be accounted for by some combination of these
processes.

Now let's become quantitative. The energy transported by a
train of continuous waves is usually described in terms of its
intensity I, defined as the energy flux per second through a unit
area of the medium oriented perpendicular to the direction of
travel. The intensity of a light beam, a sound beam, or a radio
beam, would be expressed in SI units in Wm^{-2}.

Suppose we consider a beam of sound and set about determining its intensity. The energy in such a beam is partly the kinetic
energy of motion of the layers of the medium through which the
beam is passing, and partly the potential energy of compression
and rarefaction of those layers undergoing such perturbations at
any particular instant. If we look at a volume element dV in the
beam, small enough to be essentially at the same phase throughout, we can write that the instantaneous kinetic energy of the
medium in this volume element is

$$E_k = \tfrac{1}{2}d\,dV u^2,$$

where d is the density of the medium and u is the velocity of
motion of the volume element, not the velocity v of the wave.

Equation (7.3) gives the displacement P of the layer of the
medium containing this volume element from its normal rest
position as

$$P = a\sin\left(\frac{2\pi}{\lambda}(x - vt)\right).$$

We differentiate P with respect to time to obtain the velocity

$$\frac{dP}{dt} = u = -a\frac{2\pi v}{\lambda}\cos\left(\frac{2\pi}{\lambda}(x - vt)\right).$$

Therefore the kinetic energy of the volume element must be

$$E_k = 2d\,dV a^2 \pi^2 \frac{v^2}{\lambda^2}\cos^2\left(\frac{2\pi}{\lambda}(x - vt)\right).$$

At those instants during the cycle at which the kinetic energy is
maximum, that is, when the volume element is passing through
the center of its periodic motion, the energy is entirely kinetic.
These instants occur when the $\cos^2$ factor above is unity and

$$E_{k,max} = E_{tot} = 2d\,dV a^2 \pi^2 v^2/\lambda^2.$$

The total energy per unit volume is therefore

$$E_{tot} = 2da^2\pi^2 v^2/\lambda^2.$$

We can substitute the frequency f for the quantity v/λ obtaining

$$E_{tot} = 2da^2\pi^2 f^2.$$

Finally, the energy flux through or incident upon a surface of unit area perpendicular to the beam will be the product of the energy per unit volume and the velocity of the waves in the beam, thus

$$\boxed{I = 2da^2\pi^2 f^2 v.} \tag{7.11}$$

This tells us that the intensity of the sound is proportional to the *square* of the amplitude of the waves. This result is quite general; it holds for waves of all kinds. In electricity, for example, the power delivered to a resistive load by an AC electrical wave arriving on a transmission line is proportional to the square of the RMS voltage or the RMS current.

7.4 Reflection

We have already mentioned the reflection of waves, but the subject of reflection includes such a gold mine of insight as to warrant a separate section. We define reflection as the turning back of a wave by some obstacle in its propagation path. Reflection takes place only where there is an abrupt change of the transmission characteristics of the medium in which the wave is traveling. Conversely, wherever such an abrupt change exists, reflection always occurs. The reflection may be partial or total, depending on the nature of the discontinuity causing it. For example, a beam of light passing from air into glass is partly reflected at the air/glass boundary and partly transmitted into the glass. There is no energy absorption at the boundary. The same beam of light encountering a metal mirror surface will be almost totally reflected. If the light is beamed instead into a tank of muddy water, it will be partly reflected and partly transmitted; however, the transmitted portion will be progressively absorbed exponentially as it penetrates into the liquid. A wave of voltage traveling along a transmission line may be totally absorbed by a motor doing useful work connected at the end of the line. On the other hand, if the motor is idling, most of the wave energy is reflected back onto the line. As you can see from these examples, reflection is a many-splendored thing. We shall consider it in more detail in Chapter 12.

► **Exercises**

7.1   From your own observation, what evidence can you suggest to verify that the speed of sound is the same for all frequencies?

7.2   If a single disturbance sends out both transverse and longitudinal waves that travel with known different speeds in the medium, how could the origin of the disturbance be located? Describe how this method could be used to locate sources of earthquakes?

7.3   How could you show by an experiment that energy is associated with a wave?

7.4   A tuning fork is set in vibration and its base is held against a small open wooden box. What effect does the box have on (a) the loudness, and (b) the duration of the sound?

7.5   Show that the velocity of sound in an ideal gas is independent of the gas pressure.

7.6   How will the speed of waves on a string be changed if you both double the tension and halve the mass per unit length?

7.7   Show why for very intense sound waves in air the condensed portions of the waves tend to over-run the rarefied portions.

7.8   For waves in a solid rod, show that no available tension will make transverse waves go as fast as longitudinal waves.

7.9   Write an equation to describe each of these progressive waves: (a) amplitude 0.20 cm, period 0.02 s, wavelength 500 cm; and (b) amplitude 1.25×10^{-7} cm, frequency 256 Hz, speed of propagation 330 m s^{-1}.

7.10   A sine wave traveling to the right in a cord is represented at time $t = 0$ by the full curve in figure 7.6. The broken curve represents the shape of the cord at time $t = 0.12$ s. Find (a) the wavelength, (b) the amplitude, (c) the speed, (d) the frequency, and (e) the period of this wave.

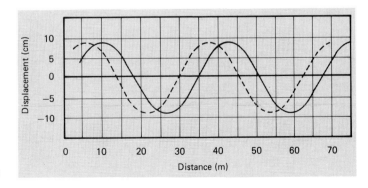

Figure 7.6

7.11   Show by differentiation and substitution that equation (7.2) is a solution of the wave equation.

7.12   A wave is represented by the equation $y = 0.20 \sin [0.40\pi (x - 60t)]$, where distances are measured in centimeters and time in seconds. Find (a) the amplitude, (b) the wavelength, (c) the speed, and (d) the frequency of this wave. (e) What is the displacement at $x = 5.5$ cm and $t = 0.20$ s?

7.13   Compute the intensity of a compressional wave in air, at $0°C$ and 760 mm Hg, if its frequency is 1056 Hz and its amplitude is 0.00120 cm. The speed of the wave in air is 331 m s^{-1}.

7.14   Show that the energy density in a 1000 Hz sound beam having particle displacement amplitude 1.50×10^{-5} m is 0.0057 J m^{-3}.

7.15   Show that a wave having a shape $P = a \cos^2 (x - vt)$ is also a solution of the wave equation.

8 Standing Waves and Resonance

Here we are concerned with one of the most ancient branches of mathematics, the theory of the vibrating string, which has its roots in the ideas of the Greek mathematician Pythagoras.

Norbert Wiener

8.1 Formation of Standing Waves

In the last chapter we dealt with some of the common features of the behavior of running waves. In this chapter we shall discuss two other properties exhibited by wave systems of all kinds: standing waves and resonance.

To be specific, let us consider a train of running waves of transverse displacement traveling from left to right in the positive x directon along a string. Let the string be terminated at $x = 0$, where it is clamped to a rigid support. Transverse forces can be brought to bear upon this support by the arriving waves, but, if it is truly rigid, the support will not move. Thus there is no mechanism whereby energy can be absorbed by the support; the oncoming waves will be totally reflected. There will then be two trains of waves traveling in opposite directions on the string ahead of the support: the orginal train going from left to right and the reflected train going from right to left. These wave trains will move through each other. They have the same velocity, the same frequency and wavelength, the same amplitude, but opposite directions of travel.

The equation describing this situation will include both terms from the right-hand side of equation (7.2):

$$P = f_1(x - vt) + f_2(x + vt),$$

where the wave shape function f_2 is the mirror image of function f_1. If we specify that the waves are sinusoidal in shape, the functions f_1 and f_2 are identical since the sine function is its own mirror image. Then

$$P = a \sin\left[2\pi(x - vt)/\lambda\right] + a \sin\left[2\pi(x + vt)/\lambda\right].$$

These wave trains travel through each other, producing by their superposition a resultant which can be determined by expanding the terms of this expression:

$$P = a \sin(2\pi x/\lambda) \cos(2\pi vt/\lambda) - a \cos(2\pi x/\lambda) \sin(2\pi vt/\lambda)$$
$$+ a \sin(2\pi x/\lambda) \cos(2\pi vt/\lambda) + a \cos(2\pi x/\lambda) \sin(2\pi vt/\lambda),$$

$$\boxed{P = 2a \sin(2\pi x/\lambda) \cos(2\pi vt/\lambda).} \qquad (8.1)$$

This equation tells us that the resultant wave pattern at any instant has a sinusoidal shape given by the factor $\sin(2\pi x/\lambda)$ with twice the amplitude of the original wave. Further, this sinusoidal displacement fluctuates periodically with time from amplitude $2a$ through zero to $-2a$, then back through zero to $2a$, according to the cosine factor, $\cos(2\pi vt/\lambda)$. Even though this wave pattern was synthesized from running wave trains going in opposite directions, you see no appearance of running waves when you look at it. Various segments of the wave medium merely bob up and down without seeming to go anywhere. Equation (8.1) defines what is called a standing wave. Figure 8.1 shows a family of plots of equation (8.1) for several successive instants of time.

The following features of this standing wave are worthy of note. Firstly, setting $x = 0$ in equation (8.1), we see that the clamp is a point of zero displacement for all values of time. This is not surprising, since the clamp, by definition, enforces zero displacement at the place where the clamp is. Secondly, additional points of zero displacement are found elsewhere along the string. These 'dead spots' are called *nodes*. Their locations can be deduced mathematically by enquiring for what values of x the shape factor $\sin(2\pi x/\lambda)$ is zero. Thus, whenever $2a \sin(2\pi x/\lambda) = 0$, x must have the value $\pm n\lambda/2$, where n is any whole number. The nodes are seen to occur every half wavelength along the string, beginning with the one at the clamp. The segments of the string between the nodes are often called *loops*. If the string is vibrating at a frequency higher than that which the eye can follow, the successive instantaneous shapes assumed by the string will merge into a blur, so that one sees simply a number of vibrating segments separated by the nodes. Thirdly, the time-variation factor $\cos(2\pi vt/\lambda)$ shows that the amplitude of the sine-wave shape of the string varies sinusoidally with time with the same frequency v/λ as the frequency of the original running waves. These features of standing waves are common to wave systems of all kinds, as will be apparent from the following comparisons.

8.2 Two-wire Transmission Line

An open circuit at the far end of an AC transmission line is the electrical analog of a mechanical clamp at the end of a string. As a

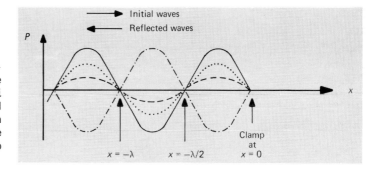

Figure 8.1 A standing wave is a function of both distance and time. The shape of the wave varies according to $\sin 2\pi x/\lambda$ and is shown for three values of t: full curve, $t = 0$; dotted curve, $t = t_1$; broken curve, $t = t_2$. The amplitude of the sine wave (chain curve) varies according to $\cos 2\pi vt/\lambda$.

termination on the vibrating string, the clamp prevents transverse motion. Similarly, in the electrical analog the open circuit termination prevents the flow of charge between the wires and causes waves of current and voltage to be reflected back along the transmission line towards the generator end. The returning waves have amplitude undiminished by the reflection, since there is no energy-absorbing mechanism associated with the open circuit. They also have the same frequency as the generator, and therefore fulfill the condition for a standing wave, that there be a second wave train traveling in the opposite direction and having the same frequency and amplitude as the original waves. The standing wave of current will have its first node at the open-circuit end, with other nodes every half wavelength from there back towards the input end of the line.

At the same time a standing wave of voltage likewise occurs, with the first node a quarter wavelength in front of the open-circuit end and every half wavelength from there backward. The nodes of current and the nodes of voltage interleave each other; each node of voltage is bracketed by two nodes of current a quarter of a wavelength away in either direction. Likewise nodes of current are bracketed by nodes of voltage.

▶ **Example 8.1** The speed of propagation of 400 Hz waves on a certain two-wire power transmission line is one tenth the speed of light in free space. In a storm a tree is blown over onto the line, breaking the wires. A serviceman traveling along the line to locate the fault observes, by connecting a voltmeter between the wires at various places, that there is a node of voltage at a distance of 16.3 miles from the power station. With this information, what inference can he make as to the location of the fault?

In this example $v = 18\,650$ mile s^{-1} and $f = 400$ Hz. Frequency, λ, is given by

$$\lambda = \frac{v}{f},$$

$$= \frac{18\,650}{400},$$

$$= 46.6 \text{ mile}.$$

A current node would appear at the break and a voltage node at a quarter of a wavelength away from it. Other voltage nodes would be half a wavelength apart. Therefore the break must be 11.7, 35.0, 58.3, . . . miles from the node located by the serviceman, or at a total distance of 28.0, 51.3, 74.6, . . . miles from the power station.

8.3 Sound Waves in a Tube

Consider a continuous train of sound waves traveling along the interior of a tube terminated at the far end by a rigid closure. Such a termination is analogous to the clamped end of a vibrating string, in that the closure prevents the longitudinal motion of the adjacent layer of air. The sound is totally reflected, and the reflected wave train proceeds in the opposite direction, setting up a standing wave in the tube as it interacts with the oncoming wave train. The termination itself becomes a node of longitudinal motion, with other nodes disposed half a wavelength apart within the tube. Interleaved with the nodes of motion are nodes of pressure, also half a wavelength apart from each other and a quarter wavelength from the bracketing nodes of motion.

8.4 Beam Incident upon a Reflector

Consider a microwave beam incident upon an ideally reflecting metal mirror surface. As the alternating electric field of the microwave penetrates the surface layer of the mirror, it encounters a sea of free electrons in the metal. If the electric field is polarized vertically, the electrons will oscillate up and down as they are accelerated by the field. When charges are accelerated they radiate electromagnetic energy. The surface layer of the mirror containing these oscillating electrons thus constitutes an antenna which re-radiates the incoming energy backwards along the direction of the incident beam. If the metal is of low electrical resistivity, the reflection can be almost total. Since the electrons are oscillating with the frequency of the incoming microwaves, the reflected microwaves will also have that frequency. The condition for a standing wave is thus fulfilled, and the space in front of the mirror will contain a standing electromagnetic wave.

Another way of looking at this situation is to regard the free electrons in the metal surface layer as a short-circuiting termination of the microwave beam. If the electrons were completely free, the reflection would be total. However, the moving electrons constitute a current, and all metals exhibit some resistance to current. A small portion of the incident wave energy is therefore converted to Joule heating in the surface layer of the reflector and the remainder, perhaps 98 %, is reflected.

The standing wave in front of the mirror can be analyzed with a detector, and a succession of voltage nodes spaced half a wavelength apart along the axis of the beam will be found.

Interleaved with the voltage nodes, but not picked up by the usual diode detector, are nodes of displacement current.

These specific examples, chosen from various fields of physics and engineering, should impress you further that waves of all kinds do indeed behave alike. If you can learn wave physics in generalized terms, as we have attempted to develop it in this book, you need not search the literature for a specific treatment of a particular wave system. You need only to take a generalized solution and fit into it the particular symbols appropriate to the kind of wave system you are dealing with.

8.5 Resonance in Wave Systems

To establish further the universality of the basic wave concepts, let us turn to the consideration of another aspect of wave behavior which pervades the length and breadth of wave physics: resonance.

We brought up resonance in Chapter 6 where we described the properties of oscillators. There we showed that when a system capable of periodic vibration is driven at a frequency coinciding with the natural frequency of the system, the amplitude of the vibration becomes larger than for any other driving frequency. Thus, a child's swing describes a large arc only when it is driven by pushes which coincide with the natural period of the swing. The tuning circuit of a radio receiver will respond only when it is exposed to incoming radio waves of the same frequency as that at which it is set to oscillate normally.

A wave system is also an oscillator. To pursue this line of thought, let's return to our vibrating string with a terminating clamp. In § 8.1 we considered waves coming along the string and being reflected at the termination to produce a standing wave, but we said nothing about where those waves were coming from or how they initially were being produced. Let's suppose that the waves are being generated at the input end of the string by the vibrating prong of an electrically driven tuning fork to which the end of the string is attached. Now we have to recognize that the tuning fork is not only the generator of new waves, but also the reflector of previously generated waves that have already made the round trip to the clamped end and back. If we could concentrate our attention on a single wave crest as soon as it is generated and launched onto the string, we would follow it along to the clamp, witness its reflection there, and return with it to the generator, where it would be reflected again by the tuning fork, and so on. The crest would be reflected repeatedly at both ends of the string and would make several back-and-forth journeys before finally dying out because of the various energy-absorbing processes occurring on the string, principally the radiation of sound. In making these excursions along the string our wave would, of course, become indistinguishably mixed up with other waves doing the same thing and create the standing wave which would result. All this time new waves would be generated and fed onto the string by the tuning fork.

Now—and here's the way we arrive at resonance—suppose the length of the string is such that a wave crest takes exactly a whole number of wave periods to make the round trip and arrive back at the starting point for its second journey along the string. It will begin this second journey just at the instant the tuning fork is launching a new wave. This new wave will superpose upon the already existing wave, crest on crest, to give a wave with nearly double the amplitude of the former crest alone. After another round trip to the clamp and back, the combined crest will receive further augmentation by the addition of another wave crest by the generator. The amplitude thus continues to build up until the energy losses during each excursion (losses that increase as the amplitude increases) become equal to the new energy fed into the system during each cycle of the generator. The final result, after several transits of the string and the establishment of a steady-state amplitude, is a standing wave of abnormally large amplitude.

In order for this condition of resonance to occur, the generation of new wave crests must coincide exactly with the reflected return of previously generated wave crests. Fulfillment of this condition requires that the round-trip travel time be an exact whole number n of wave periods, that is $2l/v = nT$, where l is the length of the string and T the period. This condition, in turn, requires that the length of the string be an integral number of half wavelengths, $l = \frac{1}{2}n\lambda$, i.e. the string must contain an exact whole number of loops of the standing wave.

The process whereby the dimensions of the wave medium are adjusted as described above to produce resonance is called *tuning*. Alternatively, a wave system with fixed dimensions can achieve resonance if we adjust the frequency of the source, and hence the wavelength, so that the condition $l = \frac{1}{2}n\lambda$ is fulfilled.

Each of the oscillators we considered in Chapter 6 exhibited a single natural frequency. Wave-system oscillators, on the other hand, exhibit a spectrum of different natural frequencies and can therefore be made to resonate at any desired one of those frequencies by suitable adjustment of the generator frequency.

For example, the string can vibrate naturally as a single loop at a frequency given by $v/2l$. It also has natural vibration frequencies at all integral multiples $nv/2l$ of this base frequency. If one could slowly increase the frequency of the tuning-fork driver, one would see the standing wave on the string suddenly break into resonance as each one of the natural frequencies of the string is swept through by the driving frequency. Figure 8.2 shows this spectrum of resonances, each of which may be compared with the single-resonance curve for, say, a spring-and-mass oscillator. Electrical transmission lines, organ pipes and other wind instruments, echo chambers, and vibrating surfaces all exhibit similar multi-resonant spectra.

As a practical matter, the experiment just described can not be performed, as a tuning fork vibrates at one frequency only. However, one can set up an equivalent experiment in which one varies not the frequency of the driver, but rather the velocity of the wave on the string by varying its tension. Since $f_n = nv/2l$ and $v = \sqrt{(T_n/\mu)}$,

$$f_n = \frac{n}{2l}\sqrt{\frac{T_n}{\mu}},$$

from which we obtain:

$$T_n = \frac{f_n^2 \, 4l^2 \mu}{n^2}. \tag{8.2}$$

Hence with an electrically driven tuning fork of fixed frequency f one can excite various resonances by adjusting the tension of the string, as shown in figure 8.3.

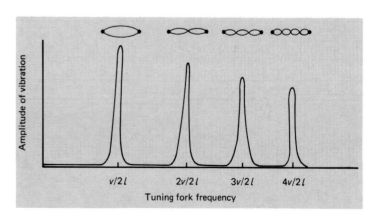

Figure 8.2 A vibrating string exhibits many resonances.

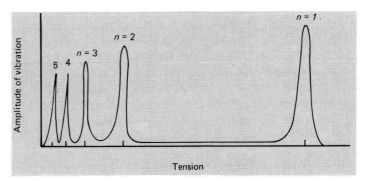

Figure 8.3 The spectrum of resonances can be revealed by varying the tension of a vibrating string.

▶ **Exercises**

8.1   Water waves of wavelength λ and speed v are reflected from a smooth vertical wall, forming a standing wave. For a buoy floating at distance x_b from the wall, find (a) the vertical displacement and (b) the vertical velocity at any instant.

8.2   Assume that you are standing with a tuning fork in your left hand and a smooth wall at your right. One ear detects no sound, the other is where it hears the sound loudest. Your ears are 18 cm apart. What are the possible frequencies for the tuning fork?

8.3   A wave having shape $P = a \sin^2[2\pi(x - vt)/\lambda]$ is reflected back onto itself. Show that the resulting standing wave has half the wavelength of the standing wave formed by reflection of a wave having shape $P = a \sin[2\pi(x - vt)/\lambda]$.

8.4   A stretched string with waves produced by an electrically driven tuning fork is found to resonate with successively higher numbers of loops when the tension is 430, 242, and 155 N. What must be the tension for the string to resonate as a single loop?

8.5   What are the periods T_n of natural oscillations in a two-conductor cable of length 0.1 m with open circuits at both ends when submerged in water (dielectric constant $\varepsilon = 80.1$)? The wave speed in water is $c/\sqrt{\varepsilon}$.

8.6   While watching a TV screen from across the room you can make horizontal lines appear on the screen by humming. By changing the pitch of your humming you may be able to make the pattern of lines move up, or down, or remain stationary. How do you explain this connection between humming and vision?

8.7   A beam of single-frequency sound is directed perpendicularly onto a hard reflecting wall. Describe a method by which you might determine the frequency of the sound.

8.8   A man carrying a short-wave radio receiver notices that as he approaches a building with a metal curtain wall the sound from the radio goes through several maxima and minima of loudness. Explain.

8.9   If a node on a vibrating string is a place of zero displacement, how does energy from a wave generator get through a node to produce loops of displacement beyond it?

8.10   In a poorly designed auditorium a man sitting in a particular seat can hear the speaker clearly while a person two seats behind him can hear the speaker hardly at all. Explain. (Assume that both listeners have equal hearing acuity.)

9 Wave Interference and Interferometry

We shall see in this chapter how sounds quarrel, fight, and when they are of equal strength destroy one another, and give place to silence.

Sir Robert Stawell Ball

9.1 Introduction

We have already considered one case of the interference of waves, in which two wave trains of the same amplitude and frequency, traveling in opposite directions, interact to produce standing waves. We saw that at the nodes of any standing wave the interaction, or interference, of the two wave trains always produces a condition of zero displacement at all phases of the wave cycle. Such annihilation is called destructive interference. At the loops, the interacting wave trains superpose to produce a motion having twice the amplitude of either wave alone. This enhancement is called constructive interference. Interference occurs when two or more waves, pulses, or wave trains interact at the same place. The results are interesting and often useful. Interferometry is the technology of deliberately producing such interactions in order to achieve some desirable end.

Wave interference exhibits a consistency described formally by the principle of superposition, which says that whenever two waves or wave trains interact, the resulting motion is the algebraic sum of the motions that would be produced by the individual waves or wave trains at that place and instant. This principle is clearly exemplified in the standing wave systems we considered in Chapter 8. We shall now consider other cases of interference and show how interferometry can be employed to assist in the making of precise measurements.

9.2 Beats

An elementary treatment of the phenomenon of beats is usually introduced by describing what you hear when you listen to two pure tones of equal amplitude and slightly different frequencies f_1 and f_2. Let the pressure fluctuations of these two sounds be specified by $P_1 = a \sin (2\pi f_1 t)$ and $P_2 = a \sin (2\pi f_2 t)$. By the principle of superposition their combined effect upon your eardrum will be:

$$P_1 + P_2 = a[\sin (2\pi f_1 t) + \sin (2\pi f_2 t)]$$
$$= 2a \sin [\tfrac{1}{2}(2\pi f_1 t + 2\pi f_2 t)] \cos [\tfrac{1}{2}(2\pi f_1 t - 2\pi f_2 t)].$$

This is the equation of a sinusoidal pressure wave having peak amplitude $2a$ and frequency $\tfrac{1}{2}(f_1 + f_2)$, the average of the two frequencies being combined. The peak amplitude $2a$ is modulated by the factor $\cos [\tfrac{1}{2}(2\pi f_1 t - 2\pi f_2 t)]$. This factor can be rewritten $\cos [2\pi (f_1 - f_2)t/2]$, suggesting that the modulation occurs at a frequency given by *half* the difference between the two frequencies. It does, but the cosine factor passes through zero at *twice* the frequency indicated by the argument. Therefore the sound appears to come in bursts at frequency $f_1 - f_2$ rather than $\frac{1}{2}(f_1 - f_2)$. Figure 9.1 shows a plot of such a modulated wave. If the original frequencies f_1 and f_2 differ by less than 8 or 10 Hz, the modulating frequency can easily be detected by the ear as an amplitude fluctuation of the sound. These amplitude fluctuations are called *beats*.

An interesting method for producing visible beats is the Moiré technique. Let a grid pattern of regularly spaced black and white lines be photographed onto a photographic plate, with the number of lines per inch given by f_1. Then let a similar pattern of lines be photographed onto another plate with f_2 lines per inch. If the plates are then superposed with the two sets of lines oriented parallel, by looking through the plates at a sheet of clear white paper one can easily see a coarser pattern of alternating brightness and darkness at $f_1 - f_2$ alternations per inch. This pattern is called a Moiré pattern. It is a beat phenomenon just as much as the beats with sound. The Moiré beats, however, are beats in space rather than beats in time. Instead of having a pressure amplitude fluctuating because of the superposition of two functions of time with different frequencies, the Moiré pattern is an intensity-amplitude fluctuation resulting from the superposition of two functions of distance with different spacings.

A Moiré pattern of fringes can also be produced with two Moiré plates having grids of identical spacing, with a slight angular inclination between the two. In this case, however, the light and dark fringes will run *crosswise* to the direction of the lines of the interfering grids. Very fascinating aesthetic fringe patterns result from other Moiré plate designs. For example, if two identical plates are prepared with a radial design like the spokes of a wheel, and their superposition is viewed with the centers slightly displaced, the fringe pattern will resemble the pattern of lines of force in the field of a magnetic dipole.

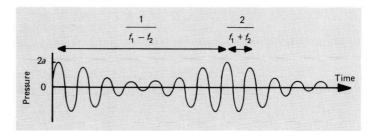

Figure 9.1 The amplitude of the two beating sounds fluctuates as the difference frequency.

The production of beats has practical uses. When a technician tunes a piano he listens for beats occurring when the piano note is sounded simultaneously with the note from his standard pitch pipe or tuning fork. He increases or decreases the tension in the piano string until no beats are heard. He then knows that the piano frequency is that of his standard. In most radio and TV receivers the incoming signal, at frequencies ranging from hundreds of kilohertz to several megahertz, is combined with a steady signal of fixed frequency generated within the receiver, to produce a lower difference frequency which is easier to process in the succeeding stages of the receiver. The process of beating one radio frequency with another is called heterodyning. It produces a beat frequency (termed intermediate frequency, or IF) which permits sharper tuning and greater station selectivity than would be possible with the simple resonant-circuit tuner of figure 6.5.

Moiré fringes are used to monitor the perfection or imperfection of the spacing of elements of periodic structure or design. Lord Rayleigh used Moiré interferometry to study the irregularities of diffraction gratings. Similar techniques are employed in the weaving industry, in engraving, and in strain measurements where the ultra-precision of optical-wave interferometry is not required.

Sometimes beats lead to undesirable situations. In music, if two notes sounded together as a chord have a difference frequency that is not a simple submultiple of either separate note, the resulting sound may be heard as a discord. Such combinations are avoided by composers and musicians who wish their music to be soothing. In airplanes and boats having two engines and two propellors, if the engines are not run synchronously their beat frequency may coincide with the natural frequency of some mechanical vibration of the structure and thus excite it to a resonance which may ultimately weaken the structure. While the pilots of such craft can hardly maintain exact synchronism, they can usually synchronize the engines closely enough so that the beat frequency is too low to excite body resonances.

9.3 Path-difference Interferometry

In order to have interference it is clear that two sources of waves are necessary. In the production of standing waves the second source is supplied by the reflector which reverses the wave train back upon itself. In the production of beats the two sources are separated and have different frequencies. Beats may therefore be regarded as an example of frequency-difference interferometry.

There is yet another type of interferometry in which the two sources are physically separated and have the same frequency but are at different distances from the collector or detector. Such interaction might be called path-difference interferometry.

Perhaps the simplest example of path-difference interferometry is pictured in figure 9.2. Two plane glass plates are held together with a thin tapering wedge of air between them. This air wedge is illuminated from above by monochromatic light from a large-area source reflected from a glass plate at 45°, as shown. Two rays of light reflected back to the eye by the two inner air/glass interfaces will superpose at the eye to produce constructive or destructive interference depending on the thickness of the air film and the effective optical path difference between the two rays.

Let's suppose that the full-line ray path in figure 9.2 leading to the two reflected interfering rays from region A of the air film results in the appearance of a bright band of light appearing at A. There will be some neighboring region B where the air film is a quarter of a wavelength thinner from which two reflected rays to the eye will interfere destructively, producing a dark band at B. A series of alternating dark and light bands will accordingly be seen as one scans the tapered air film from one end to the other. The thickness of the film at any point can be deduced from the number of the bands and the wavelength of the light. Each fringe corresponds to an air-film thickness half a wavelength more or less than the thickness at the next adjacent fringe. In this experiment the two sources of light required for interference are the reflections of the original ray in the upper and lower surfaces of the air film.

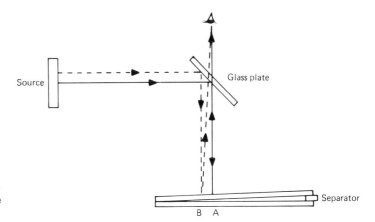

Figure 9.2 Interference fringes are produced by reflection at the surfaces of the tapered air film.

► **Example 9.1** To measure the thickness of a film of cellophane a man uses an arrangement similar to that of figure 9.2 except that the glass plates are separated by one thickness of the cellophane at one end and by two thicknesses at the other. He counts 64 bright fringes of sodium light ($\lambda = 5893$ Å) along the length of the plates. How thick is the cellophane?

Each fringe means another half-wavelength thickness of the air film. Therefore the total difference in thickness between the ends, which is the thickness of one layer of cellophane, is

$$64 \times \frac{0.00005893}{2} \text{ cm} = 0.00189 \text{ cm}.$$

Interference bands or fringes of this kind are used to test the perfection of optical surfaces. The surface of a lens undergoing test is placed in contact with a mating surface known to be perfect. If the curvature of the test lens is different from that of the standard surface, a pattern of ring-shaped interference fringes will be seen in the air film between the two surfaces. Other imperfections of the figure of the tested surface will show up as irregularities of the circular shape of the fringes. If the two surfaces match each other perfectly, no fringes will be seen; the entire area of the air film will be uniformly bright (or dark). Departures of around 1/20 of a wavelength from a perfect figure can be detected by this interferometric technique.

An interferometric device in which the two interfering sources are not obtained by reflection is shown in figure 9.3. Monochromatic light from a source S falls simultaneously on two slits A and B in a baffle. According to Huygens' principle of wave propagation, each slit then behaves like a new source emitting secondary waves of the same frequency in all directions in the space beyond the baffle. These secondary waves then fall upon a screen, where we now wish to examine the pattern of illumination resulting from the superposition of the waves from the two slits.

Figure 9.3 is drawn with wave crests at a particular instant represented by the full-line arcs centered on the two slits. The wave troughs at the same instant are represented by the broken-line arcs. It is clear from a careful inspection of this pattern that there are certain directions along which wave crests from one slit always superpose on wave troughs from the other slit. These directions are indicated by the bold lines. Along these lines destructive interference takes place; where they intersect the screen no light falls. These lines of destructive interference are called nodal lines. When the analog of this experiment is carried

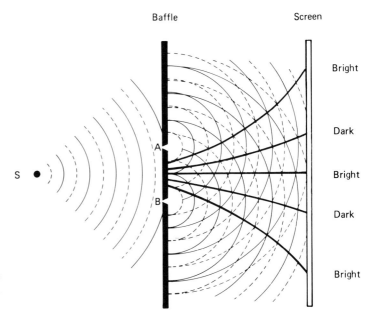

Baffle   Screen

Bright

Dark

Bright

Dark

Bright

Figure 9.3 The double-slit interference pattern consists of alternating bright and dark fringes on the screen.

out in the laboratory with water surface waves in a ripple tank, one can readily see the nodal lines on the water surface.

Between the nodal directions there are directions along which constructive interference occurs, with wave crests and troughs from both slits superposing in phase. The screen receives reinforced illumination in these directions. The pattern of illumination on the screen consists, therefore, of a series of alternating bright and dark fringes. A bright fringe occurs on the screen where the optical path difference between the two slits and that point on the screen is a whole number of wavelengths. The dark fringes occur where the path difference is an odd number of half wavelengths. Knowing the separation of the slits, the distance between the fringes on the screen, and the slit-to-screen distance, one can determine the wavelength of the light.

▶ **Example 9.2** What is the wavelength of the illuminating source in figure 9.3 if the bright fringes are 2 mm apart on a screen 1 m away from a double slit with separation 0.2 mm?

The central fringe of the intensity pattern will be bright, since it is equidistant from both slits. If figure 9.3 is redrawn with straight-line ray paths from slits A and B to the adjacent bright fringe on the screen, it can be seen that the path from A will be one wavelength longer than that from B. Therefore,

$$\frac{\lambda}{AB} = \frac{2\,mm}{1\,m},$$

$$\lambda = 0.02\frac{0.2}{100} = 0.00004\,cm.$$

An experiment of this kind performed by Thomas Young in 1801 settled a hundred-year-old argument between people who believed that light was a wave phenomenon and others who, like Isaac Newton, believed that a ray of light consisted of a stream of particles of some kind. Since it was conceded that particles could hardly produce interference fringes, the matter was resolved convincingly in favor of the wave viewpoint. The wave theory of light remained unchallenged until the twentieth-century advent of quantum physics and is still very much a part of the modern concept of wave–particle duality.

One may logically enquire what becomes of the energy that would otherwise go to the dark fringes of the interference pattern on the screen. Obviously, if the opaque sheet containing the two slits weren't present, the screen would be uniformly illuminated. The answer is that energy channeled away from the dark fringes is piled up as additional illumination in the bright fringes. At the center of a bright fringe the illumination amplitude, by super-position, is twice what it would be from a single slit alone; hence the intensity at the center of a bright fringe is *four* times what it would be from a single slit alone. The total energy integrated over the entire fringe pattern is just twice what it would be from one slit alone, but is distributed non-uniformly into the bright fringes at the expense of the illumination of the dark fringes.

An ingenious alternative to the double-slit interferometer is the Lloyd's mirror arrangement of figure 9.4. Rays of light from the source S reach the screen by direct transmission and also by reflection in the plane mirror. The reflected ray appears to come

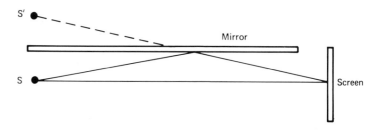

Figure 9.4 The Lloyd's mirror is a modification of the double-slit inter-ferometer.

from the source S' on a perpendicular line as far behind the mirror as S is in front. We thus have two sources of the same frequency and phase and very nearly of the same intensity. Interference fringes will therefore appear on the screen. We would expect bright fringes to occur at places where the ray paths from S and S' differ by whole numbers of wavelengths, and dark fringes to appear at places where the path difference is an odd number of half wavelengths. However, when the experiment is set up it is found that the bright fringes appear where dark fringes are expected and vice versa. This reversal occurs because the ray of light appearing to come from S' experiences a phase inversion of half a period upon reflection.

While this experiment was originally designed and performed with visible light, it is particularly suitable for performance with microwaves. The mirror can be supplied by taping a sheet of aluminum foil to a vertical wall. As the microwave receiver is moved along the line of the screen in figure 9.4, maxima and minima of received intensity are detected at positions which are convenient to locate and easy to measure.

The Rayleigh interferometer is another double-slit type of instrument, with the added feature of a focusing lens in the ray paths. This interferometer, illustrated in figure 9.5, has two slits S_1 and S_2 a few millimeters wide and about 1 cm apart. Beams of monochromatic light diverging from a single source S_0 are made parallel by a lens L_1, pass through the slits and the cells p_1 and p_2, and by means of a second lens L_2 are brought to a focus where interference fringes can be caught on a screen or viewed through an eyepiece placed beyond the focus. This instrument is used to measure refractive indices of liquids and gases.

With empty cells placed in the two light paths, the position of the central bright fringe in the interference pattern is noted. One cell is then filled with the test material and the other with a reference material of known refractive index. If the refractive indices of these materials are different, the fringes will shift. The amount of shift is a measure of the difference in index. Refractive index differences of the order of one part in 10^8 can be detected.

The Fabry–Perot interferometer represents a further step in

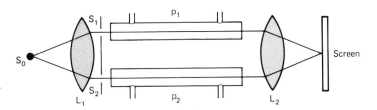

Figure 9.5 The Rayleigh interferometer uses lenses to define the light paths.

the evolution of instruments capable of extending the techniques of precision measurement. The principle of this multiple-beam interferometer is suggested by figure 9.6. Two plane glass or quartz plates are mounted precisely parallel to each other. Their facing surfaces are coated with partially transparent films of moderately high reflectivity. Light from a particular point in a large-area source S is reflected back and forth between the plates. At each reflection from the right-hand plate some of the light escapes and goes on, through the lens L, to a focus at P where it is either caught on a screen or viewed through an eyepiece placed to the right of P. The locus of all points P defined by the convergence of rays from all points of the source entering the reflector system at the same angle θ from the normal is a circle of radius OP. If the angle θ and the separation of the two mirrors is such that the path difference between rays making successive reflections is a whole number of wavelengths, a bright circular fringe will appear at this radius. With a given plate separation there will be only a few discrete angles θ for which bright circular fringes will result. If the plates are very slowly moved farther apart, the fringes will expand in diameter, while new fringes will appear in the center of the fringe system.

The Fabry–Perot interferometer has the advantage that, instead of having only two interfering sources, it has many, indicated by S', S'', etc, in figure 9.6. The effective number of such secondary sources depends on the reflectivity of the mirror surfaces. This multiplicity, combined with the fact that path differences of tens of thousands of wavelengths can be achieved with reasonable plate separations, results in fringes which are very narrow compared with the distances between them. This feature is very useful in the examination of the fine structure of spectral lines. Many spectral lines are not purely monochromatic, but consist rather of assemblages of several lines having wavelengths very close together. Such a composite line is usually seen as a single line in the typical laboratory spectroscope, which does not have the resolving power needed to separate the closely spaced components. By bringing to bear the high resolution of the

Glass plates

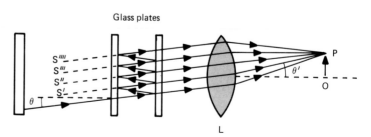

Figure 9.6 Multiple reflection of a beam occurs in a Fabry–Perot interferometer.

Fabry–Perot interferometer, one can study the detailed structure of such multiplets. This study is important because it reveals much about the internal mechanics of atomic and molecular processes.

A Michelson interferometer achieves large path differences by another method. Its basic elements are shown in figure 9.7. A ray of light from point S_1 in a large-area monochromatic source proceeds along the full line towards a glass plate b which is half-silvered (i.e., made 50% reflecting on one surface). This plate is oriented at 45° to the direction of the light path through the instrument. At the partially reflecting surface the light ray is split, half of it going to mirror m_1 where it is reflected back upon itself, the other half being transmitted to mirror m_2 where it, too, is reflected back. The two reflected beams are recombined at the half-silvered surface and proceed from there to the eye, the ray from m_1 by transmission through, that from m_2 by reflection at the half-silvered surface. If the mirrors m_1 and m_2 are exactly equidistant from the half-silvered surface, the two reflected and recombined rays will enter the eye in phase, producing the appearance of a bright field of view. If the two optical paths differ by half a wavelength, the field of view will appear dark.

Suppose now that the mirror m_2 is moved to the right by a distance of a thousand wavelengths or so. The optical paths by way of m_1 and m_2 will no longer be equal, but as long as the difference is exactly a whole number of wavelengths, the rays entering the eye along the solid line direction will still superpose constructively, giving the appearance of a bright spot in the center of the field of view. However, this time the bright center will be surrounded by a series of alternating concentric circular dark and bright fringes.

To see how these fringes are formed when the optical paths by way of m_1 and m_2 differ by a considerable number of wavelengths, let us consider another ray of light (the broken line)

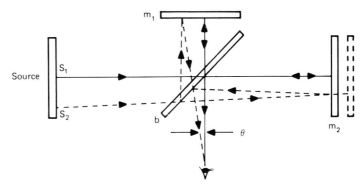

Figure 9.7 The Michelson interferometer provides for large path differences.

coming from some other point S_2 of the large-area source and reaching the eye by the broken-line paths to m_1 and m_2 inclined at an angle θ to the central ray direction. Because of this inclination, the two broken optical paths will not have the same difference as for the central ray paths, and may enter the eye out of phase with each other by half a period. In this case a dark circular fringe will appear surrounding the central bright spot. This, in turn, will be surrounded by a bright fringe of inclination 2θ, and so on. The field of view will be filled with these concentric light and dark fringes.

If one of the mirrors is slightly inclined about a vertical axis, the center of the circular fringe system can be shifted out of the field of view. What one then sees is a series of vertical light and dark fringes which are arcs of circles. With mirror m_2 mounted on a carriage which can be moved back and forth by a calibrated screw, this interferometer can be used for the precise measurement of wavelengths. Alternatively, it can be used to calibrate distances in terms of wavelengths of light. Every movement of m_2 by half a wavelength results in the displacement of the fringe pattern by one fringe.

If the source emits light of more than one wavelength, each produces its own fringe pattern. The instrument can therefore be applied to the study of the fine structure of compound spectral lines. Fringes cannot ordinarily be seen with a white-light source because the fringe patterns of the different wavelengths overlap so that one sees a uniformly illuminated field of view. However, if mirror m_2 is positioned so that the path difference between the rays entering the eye from m_1 and m_2 is zero, then half a dozen fringes are seen on either side of the zeroth fringe. These fringes become more highly colored the farther away they are from the zeroth fringe, until they finally become indistinguishable because of the overlap of the fringe patterns of the different colors.

In 1887 Michelson employed the split-beam principle of his interferometer in the construction of the apparatus which he and Edward W Morley used in the famous series of ether-drift experiments. The object of this painstaking project was to determine whether the apparent velocity of light was modified in any way by the Earth's motion through the ether. The results were negative, leading to the conclusion that no matter how fast an observer is moving, or in which direction, the speed of light measured by him will always be the same. This conclusion was at odds with the expectations of that day and contradicted the results of similar experiments with sound in moving media. The Michelson–Morley experiment shook the edifice of physics. It took several decades for physicists to accept and feel easy with its null result. Even after Einstein published his special theory of relativity in 1905, the feeling persisted that nature wasn't behaving, and in subsequent years the ether-drift experiment was repeated several times until lingering doubts were laid to rest. In 1955 it was finally accepted that it is impossible to detect any 'absolute' motion of the earth in an optical experiment.

In another classical example of the usefulness of the Michelson interferometer, Michelson in 1893 determined the length of the international standard meter bar in terms of the wavelength of the strong red line in the cadmium spectrum. In 1960 the General Conference on Weights and Measures adopted a more purely monochromatic source, the yellow-orange line of the krypton-86 spectrum and defined the standard meter as 1 650 763.73 of these wavelengths.

In the interferometers we have described you may have noted that the two or more interfering rays are always derived from the same primary source, either directly or by imaging the primary source. You may ask 'Why not have two separate sources; why bother with images?' The answer is that stationary interference patterns will not result unless the two effective sources are in phase or maintain a constant phase difference over thousands or tens of thousands of wave periods. This condition is impossible to fulfill with two or more independent sources each containing many atoms radiating with random phases. However, radiation proceeding from an image always has a constant phase relationship with radiation coming from the primary source or from any other image of the primary source. The interfering rays under these conditions are said to be *coherent*.

9.4 Holography

A very special case of path-difference interferometry is the holographic process—a method of producing images by recording an interference pattern on a photographic plate and then, from this pattern, reconstructing an image of the object. To produce a hologram one must have available a light source that is coherent, that is, a source that projects a beam of light in which all the individual wave trains passing through a fixed cross section of the beam are in the same phase at the same instant (figure 9.8).

The arrangement for making a hologram is shown in figure 9.9 (*a*). Light from a coherent source—a laser—is projected through a diverging lens to give the beam enough width to illuminate an object completely. This beam is split by a partially reflecting mirror surface. One part of the split beam falls upon the object to

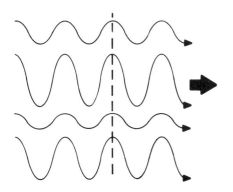

Figure 9.8 In a coherent beam all the individual wave trains are in step.

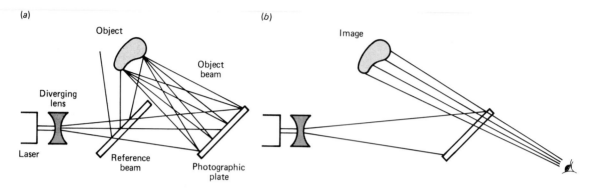

Figure 9.9 (a) A hologram is made; (b) the image is viewed.

be imaged. Light scattered and reflected from all parts of the object falls upon all parts of the photographic plate. The other part of the split beam falls directly onto the photographic plate where it interferes, constructively and destructively at different points in the emulsion, with the rays coming from the object. The plate is thus exposed and developed. We emphasize that what is captured by the emulsion is not a *picture*, but rather an *interference pattern*. This pattern has no resemblance to the object producing it.

If the developed holographic plate is now illuminated with coherent light of the same wavelength used to produce it and viewed from the opposite side, a three-dimensional virtual image of the original object will appear in the position occupied by the object at the time of exposure (figure 9.9*b*). This holographic image has an extraordinary property: if the photographic plate is broken and only a small piece retained, this small piece will contain all the information necessary to resynthesize the *entire* image. This property is in contrast with the limitation of an ordinary camera photograph. After all, if you tear an ordinary photograph in half and throw half away, you have lost half of the picture forever.

9.5 Concluding Remarks

In this chapter we have come nowhere near exhausting the field of interferometry. We have given enough examples, however, to illustrate the enormous versatility of interferometric techniques and to describe the physical processes on which these techniques are based. If you are interested in pursuing this subject more deeply, we invite you to consider the list of suggestions for further reading. You will come across other optical interferometer arrangements as devised by Mach and Zehnder, by Twyman and

Green, by Lummer and Gehrcke, by Kosters, and by Jamin. You will read about interferometric methods for measuring the diameters of stars, for gauging minute strains in deformable construction members, and for determining patterns of fluid flow. You may learn about the use of sound-wave holography for imaging the internal organs of the body, and you may discover that interferometric methods are used for obtaining high-resolution radio pictures of portions of the heavens in radio-astronomy. Blind people develop skill in the use of acoustic interferometry in establishing the direction, distance, size and shape of obstacles, and strain-gauge interferometry with laser beams is employed to detect the slight deformation of the Earth's crust that may help us to predict earthquakes. Altogether these comprise a very impressive list of applications.

▶ **Exercises**

9.1   A locomotive engineer sounds his horn as his train approaches a large building down track. He hears eight beats per second as the sound combines with the sound of the returning echo. If the horn sounds at 200 Hz, use Doppler's principle to find how fast the train is moving.

9.2   Incoming ocean waves pass two vertical columns 500 m offshore and in a line parallel to the shore. At the beach the waves produce interference 'fringes' 22.5 m apart. If the wavelength is 1.5 m, show that the columns are 33 m apart.

9.3   If one of the slits of the double-slit plate of figure 9.3 is covered with a film of transparent material, the central fringe of the interference pattern is seen to shift by 2.2 fringes' distance. The refractive index of the material is 1.4 and the wavelength of the source is 5×10^{-5} cm. Show that the thickness of the film is 0.000275 cm.

9.4   A Lloyd's mirror experiment is performed with a microwave source of 1 cm wavelength, 20 cm above the table. The mirror is a sheet of metal lying flat on the table top. A microwave receiver 3 m away is slowly raised from table-top level to a height of 50 cm. How many maxima of received intensity does it go through?

9.5   A closed chamber 2 cm long with transparent end plates is placed in one of the optical paths of a Michelson interferometer. As the air is flushed out of the chamber and replaced by a denser gas, an observer counts a displacement of the fringe pattern by 60 fringes of mercury arc light of wavelength 4.93×10^{-5} cm. What is the index of refraction of the gas relative to air?

9.6   An oil film of thickness 5×10^{-5} cm and index of refraction 1.45 is deposited uniformly on a glass plate of refractive index 1.55. At what

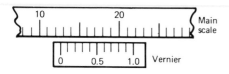

Figure 9.10 A vernier is a device for measuring subdivisions of a scale.

wavelengths in the visible region will interference in the oil film produce maximum reflected brightness when viewed normally to the film?

9.7 Consider two parallel picket fences, with boards (and spaces) 10.0 cm wide, and 10.2 cm wide respectively. As you drive past at 15 km h^{-1} and observe the Moiré pattern, how frequently can you *not* see through the two fences?

9.8 A piano tuner hears three beats per second when listening to the combined sound from his tuning fork and the piano note he is addressing. He tightens the piano string very slightly and hears five beats per second. What is his next move?

9.9 The apparatus for a double-slit experiment is completely immersed in water. How will the interference pattern on the screen be altered?

9.10 A vernier (named after Pierre Vernier $\sim$ 1630) is a short auxiliary scale placed along the main scale of an instrument to permit readings to a fraction of the least division of the main scale. The vernier is graduated in numbered divisions which are fractionally shorter (or longer) than those on the main scale. The position of the zero mark on the vernier scale between main scale divisions is read as the number on the vernier scale which lines up exactly with a graduation on the main scale. (a) Use the idea of 'beats' to justify that the reading is 12.7 on the scale shown in Figure 9.10. (b) If a circular main scale is graduated in half degrees, show that vernier divisions of 11/12 or 29/30 of the main-scale divisions will allow readings to 5 or 2 minutes of arc, respectively.

9.11 We want a fog horn to spread sound through a wide angle horizontally and waste little energy upward. Should the rectangular opening of the horn be set with its longer dimension horizontal or vertical? Why?

10 Beam and Image Formation

I have a paper afloat, with an electromagnetic theory of light, which 'til I am convinced to the contrary, I hold to be great guns.

James Clerk Maxwell (1865)

10.1 Introduction

We begin this chapter with definitions of some terms. Most of us think we know what a beam of radiation is, but let's define it anyway: a beam is a bundle of individual rays, and the rays in this bundle may be parallel, converging, or diverging as they progress. This definition requires that we now define a ray. A ray is the path followed by a very small segment of a wave front as the wave progresses. The wave front, in turn, is the locus of points of the same phase in an advancing wave. A wave crest, for example, is a wave front. Figure 10.1 depicts a beam of radiation traveling to the right, diverging from a very small-area source (ideally, a point source), with the edges of the beam defined by an aperture. The beam is made up of rays, which can be thought of as narrow pencils of radiation, everywhere perpendicular to the wave fronts. For a diverging beam, like the one shown, the wave fronts are portions of spherical surfaces expanding outwards from the source with the speed of propagation of the waves.

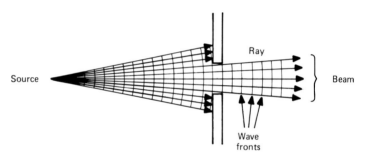

Figure 10.1 A beam of radiation is a bundle of rays.

From what has been said, it appears that a beam-defining aperture is the simplest kind of beam-forming device. This is true; apertures are used for beam formation in many applications. The beam shown in figure 10.1 is somewhat divergent. Such divergence is quite satisfactory for flashlight beams and automobile headlight beams. However, for searchlight, spotlight, and microwave communication uses, the beam should be made up of rays which are as nearly parallel as it is possible to make them. The rays of an aperture-defined beam can be made more nearly parallel simply by narrowing the aperture and thus eliminating the more highly divergent rays.† The resulting beam is described as being geometrically collimated.

---

† If the aperture is made too narrow (i.e. only a few wavelengths in width), the rays passing through it will become divergent again because of aperture diffraction (see §10.3).

10.2 Lenses

The disadvantage of geometrical collimation is the low intensity of the flux in the resulting beam. The narrower the aperture the better collimated the beam will be, that is the more nearly parallel will be the rays. However, the narrower the aperture the smaller will be the total energy in the beam.

In a lens system for light-beam formation the aperture is deliberately enlarged, and a positive lens is inserted into the aperture. If the focal length of the lens is exactly equal to the distance between the source and the lens, all light rays coming from the source and entering the lens are rendered parallel or nearly parallel in the resulting beam.

It is instructive to consider the action of a lens from the standpoint of the energy distribution on a distant target (figure 10.2). When used to produce a parallel beam, the lens gathers a bundle of rays diverging from the source S and renders them parallel. The rays are then delivered to the distant target with essentially the same energy per unit area that they had as they left the lens. After passing through the lens the energy in the beam no longer spreads out according to the inverse square law of distance, so does not get less and less intense the farther it travels. Therefore, the intensity of the radiation on the target is much greater than it would be without the lens.

▶ **Example 10.1** A searchlight lens 0.5 m in diameter forms a beam of parallel rays from a source placed 0.5 m behind the lens. How much more intense is the radiation on a target 1 km away than it would be without the lens?

Without the lens the light passing through the lens aperture would spread out 1 km away into an area that is $(1000/0.5)^2$ times larger than the lens area. With the lens in place all this radiation is concentrated at the target into an area equal to the lens area, assuming perfect collimation and no absorption or scattering loss in the air. The intensity on the target is therefore $(1000/0.5)^2 = 4 \times 10^6$ times greater with the lens present.

By placing the lens slightly farther away from the source than is required to produce a parallel beam, one can make the emerging rays *converge* so that they are focused into a small area on a target. In this case the intensity (energy per unit area) is greater still. The pattern of radiation in this intense spot is such as to form an

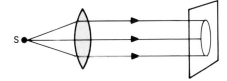

Figure 10.2 In a parallel-ray beam the intensity does not diminish with distance.

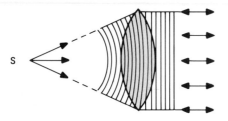

Figure 10.3 A positive lens decreases the curvature of the wave fronts.

image of the source. From the point of view of maximizing the transfer of energy from the source to the image, the lens can be regarded as a matching transformer (see §12.4).

The lens acts as it does because the refractive index of the material of which it is made (usually glass or plastic) is greater than one, and the speed of light in the material is less than it is in air. The portion of a wave front passing through the center of a lens suffers a greater retardation than those portions passing through the rim where the lens is thinner. The result (figure 10.3) is that the emerging wave front has less curvature than it had on entering the lens. If the curvature is reduced to zero, the resulting beam is parallel; if the curvature is reversed, the resulting beam converges to a focus. The change in curvature is the same, regardless of whether the beam passes through the lens from left to right, as in figure 10.3, or from right to left. The principle of reversibility recognizes the fact that a ray will follow the same path through an optical system in either direction. Therefore if the parallel rays on the right of the lens in figure 10.3 could be reversed, they would retrace their trajectories through the lens and converge to a focus at S. Beam formation and the production of images are reciprocal processes.

10.3 Lens Defects

We might describe an ideal lens as one capable of forming bright, sharp, distortion-free images of all objects within an appropriate angular field of view. No real lens or lens system attains this ideal completely because of inevitable optical aberrations. Some arise from dispersion of light within the lens, others from the geometry of rays refracted by spherical surfaces, still others from the wave nature of the radiation itself.

Among the more serious of these lens shortcomings is chromatic aberration, a defect that is present to some extent in all optical lenses and becomes limiting in microwave lenses. The index of refraction of all lens materials varies with wavelength. This property is called dispersion (see Chapter 13). It is welcome in spectroscope prisms but not in lenses. Since the focal length of a lens depends on the index of refraction of the lens material, radiations of different wavelengths are brought to a focus at different distances from the lens (figure 10.4). Chromatic aberration can be reduced to an acceptable level by making the lens in two sections of different materials having the properties described in figure 10.5. All high-quality lenses—telescope and

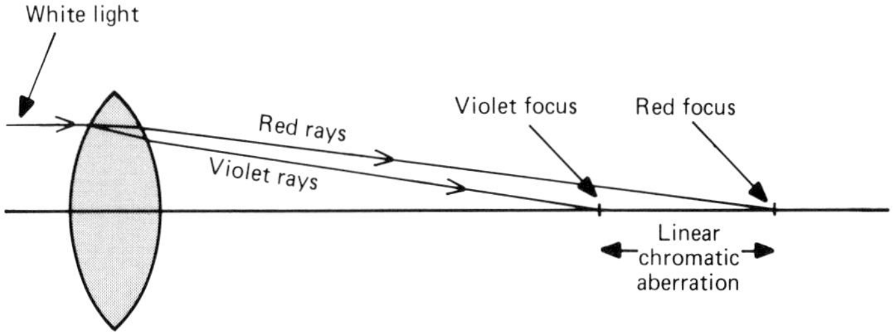

Figure 10.4 Since the index of refraction varies with the frequency of the radiation, when white light passes through a lens different colors will come to focus at slightly different points, causing chromatic aberration. (From McKelvey and Grotch, *Physics for Science and Engineering*, McGraw-Hill 1978.)

White light

Red rays

Violet rays

Violet focus   Red focus

Linear chromatic aberration

White light

Common focus

Red and violet rays

Moderate refractive index, large dispersion

Large refractive index, moderate dispersion

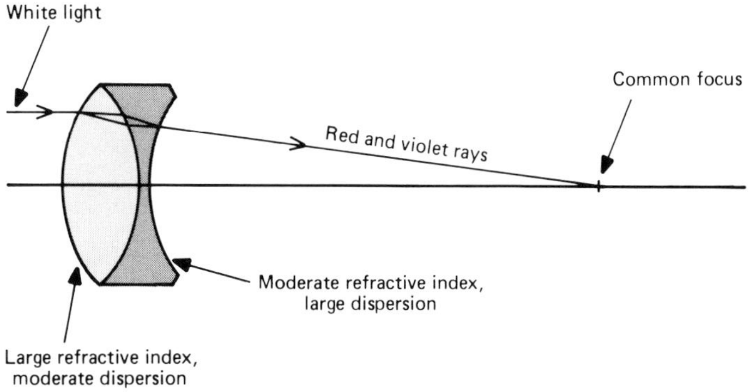

Figure 10.5 In this achromatic doublet, the dispersion in the negative (concave) lens cancels that of the positive lens without completely canceling the convergence. (From McKelvey and Grotch, *Physics for Science and Engineering*, McGraw-Hill 1978.)

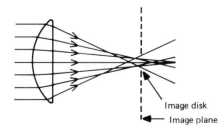

Image disk

Image plane

Figure 10.6 After passing through a simple lens initially parallel rays are not all focused at a single point.

binocular lenses, camera lenses, microscope objectives—are made in the form of such achromatic doublets.

Most lenses are made with spherical surfaces, as these are fairly easy to grind and polish. Aspherical surfaces are much harder to produce. Even with monochromatic light spherical aberration (a defect of lenses having spherical surfaces) arises from the fact that rays from an object on the optical axis after passing through the lens do not all return to the optical axis at the same point, or parallel rays (whose source is at infinity) are not brought to a single focus (figure 10.6). The same effect occurs for a mirror with spherical curvature. The simplest way to minimize spherical aberration is to use a small aperture. When the path difference for axial and marginal rays is reduced, the blurring effect of aberration is reduced. In special cases spherical aberration is avoided by using lenses with paraboloidal surfaces.

Other aberrations (astigmatism and coma) are due to the failure of parallel rays entering the lens at an angle of inclination to the optical axis to be brought precisely to a point image. Astigmatism and coma are particularly damaging to the performance of wide-angle camera lenses where much of the image is formed at large off-axis angles. These defects can be greatly reduced by using two or more separate achromatic doublets in an optical train to form a compound multiplet lens. Such lenses are more expensive to produce than simple lenses because of the additional optical surfaces which must be generated.

Even if a lens could be produced completely free of all the defects mentioned above, there would still be one more short-coming, having nothing to do with the lens itself but dependent solely upon the size of the lens opening and the wavelength of the radiation. This ultimate defect is aperture diffraction. It is unavoidable, and is present in *all* optical systems. It occurs because of the wave nature of radiation; even for a beam which, by geometrical optics, consists of strictly parallel rays, there will be a wave optical divergence of angle approximately equal to λ/D, where D is the diameter of the lens. To form a beam whose angle of divergence is only one degree, the lens must have a diameter of about 120 wavelengths. In the formation of camera images this aperture diffraction is not bothersome, as a typical camera lens may have an aperture diameter of tens of thousands of wave-lengths. It does, however, set a limit to the resolving power of telescopes and microscopes. In the microwave frequency range, because the wavelengths are of the order of centimeters, the production of reasonably parallel beams requires lenses having diameters of several meters. While glass and plastic lenses could be used to form microwave beams, size, weight, and cost make such materials impractical. Instead, microwave engineers use other beam-forming schemes.

10.4 Microwave Beam Formation by Lenses

To avoid the disadvantages of solid lenses of the size required in microwave technology, beam-forming lenses are replaced by lens-shaped structures of light styrofoam in which are distributed metal spheres or rods in a regular three-dimensional array (figure 10.7). These metal inclusions are forced into electrical oscillation by the alternating field of the radiation passing through the structure. They re-radiate in all directions, and the secondary waves thus produced interact with the primary waves to cause a progressive retardation of phase as the waves proceed through the lens. In this process the metal inclusions behave the same way at microwave frequencies as do the molecules of glass in ordinary refraction at visible frequencies (see Chapter 13). In this way microwave delay lenses having effective indices of refraction of 10 or more can be constructed.

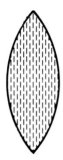

Figure 10.7 This delay lens is a three-dimensional assembly of secondary radiators.

▶ **Example 10.2** A symmetrical biconvex microwave delay lens has an effective index of refraction of 12. A microwave source is to be placed 2.0 m away from the lens on its axis. What must be the radius of curvature of the lens surfaces to produce a beam of parallel rays?

The lens maker's formula is

$$\frac{1}{f} = (n-1)\left(\frac{1}{r_1} - \frac{1}{r_2}\right),$$

where f is the focal length of the lens, n is its index of refraction, and r_1 and r_2 are the radii of curvature of its surfaces. Upon substitution, and taking the two radii of equal magnitude and opposite sign, we have

$$\frac{1}{2} = (12-1)\left(\frac{1}{r} + \frac{1}{r}\right),$$

from which $r = 44$ m.

An alternative design is the lattice or louvre lens (see figure 10.8). It consists of an array of parallel, equally-spaced flat metal plates oriented parallel to the direction of the central ray and parallel to the electric field of the microwaves. The plates are cut to such a shape that the envelope contour of the assembly is that of a *concave* lens. A glass lens of this shape would have a diverging effect on a beam of light passing through it, but for microwaves the louvre lens is *converging*.

To understand why this is so, we must investigate the behavior of a train of electromagnetic waves passing through the space between parallel metal plates separated by a distance somewhat larger than half a wavelength. Figure 10.9 shows a 'top' view of such a situation. Here the electric field of the radiation is oriented perpendicular to the page.

Upon entering the confined space the original wave train splits into two wave trains A and B which proceed with normal velocity down the alley, by multiple reflection from the inside wall of the plates. The angle between these wave trains is such that at the plate surfaces the electric fields of the two wave trains always cancel, since an electric field cannot exist tangential to a metal surface. This cancellation is indicated in figure 10.9, where the full lines represent wave crests and the broken lines represent wave troughs. At the plate surfaces the crests of wave train A are neutralized by the troughs of wave train B. The alternating electric field is strongest along the mid-line of the space between the plates, and is always zero at the surfaces of the plates.

Now concentrate your attention upon the points marked X in figure 10.9. They represent composite crests formed by the

Figure 10.8 This microwave louvre lens is made of parallel metal plates.

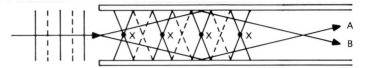

Figure 10.9 Composite crests X are formed by waves passing between parallel metal plates.

superposition of crests from wave trains A and B. As the two wave trains zigzag back and forth these supercrests move straight to the right down the mid-line of the alley. However, observe that the supercrests are farther apart than the wavelength of the original wave train. Since the composite supercrests must obviously have the same frequency as the original waves, it follows that their *velocity* must be *greater* than the velocity of the waves in free space! The term 'phase velocity' has been used to describe such a situation; the supercrests are the crests of phase waves.

'How,' you may ask, 'can these phase waves have velocities greater than the velocity of light? Isn't that the limiting velocity for anything and everything?' One of the postulates of relativity states that no material object can acquire a speed, relative to any frame of reference, greater than the speed of light, but the supercrests of figure 10.9 are not material objects; they are *conditions* and are not subject to the limiting speed of relativity.

To make this concept clearer, consider the case of two single wave crests traveling at ordinary wave speed in the directions indicated by the arrows in figure 10.10. As these waves progress, it is clear that the locus of their superposition, X, must move directly to the right with a phase velocity much higher than the normal wave velocity.

As the phase waves reach the end of the alley between the confining plates they resynthesize the train of plane waves moving away to the right with normal velocity. Now, returning to the microwave louvre structure of figure 10.8, we see that it acts like a lens having an index of refraction *less* than one. Its overall concave shape, therefore, will have an effect on radiation passing through it opposite to the effect of a lens having the same shape but with an index of refraction greater than one. The peripheral rays of the incident beam, traveling through a greater distance between the louvres, are advanced more than the central rays traveling a shorter distance through the center of the lens. The result is an emerging beam whose wave fronts have been flattened out and whose rays therefore form a parallel beam.

Comparing the modes of operation of convex glass lenses and concave louvre lenses, we observe that the glass lens reduces curvature of the wave fronts by *retarding* the central rays with respect to the peripheral rays, while the louvre lens reduces the

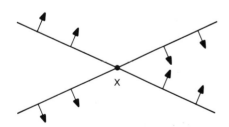

Figure 10.10 As the waves proceed, their point of superposition moves to the right with greater speed.

curvature by selectively *advancing* the peripheral rays with respect to the central rays.

From a careful study of figure 10.9 it will appear that the zigzag angle made by the ray paths A and B with the walls of the louvres must depend on the wavelength of the radiation passing through the lens. The phase velocity of the phase waves depends, therefore, on the wavelength, and so also must the effective index of refraction of the lens. The lens is thus highly chromatic and has different focal lengths for different frequencies. It cannot be used effectively for the production of well collimated microwave beams which include wide spectra of frequencies.

10.5 Zone Plates

The zone plate is a beam-forming and focusing device which depends for its operation on the constructive and destructive interference of rays traversing the structure by different paths. A zone plate for 3 cm microwaves is shown in figure 10.11. It consists of a series of concentric circular zones of radii proportional to the square roots of the whole numbers 1, 2, 3, 4, ... Every other zone between the circles is open for the transmission of the microwaves, while the zones between these are blocked with metal sheeting or mesh. The open zones are so located that the ray path from the source to each successive open zone is one wavelength longer than the path to the next smaller open zone. Portions of the incident spherical wave fronts emerging through these open zones are therefore all in the same phase and recombine on the other side of the zone plate to give plane wave fronts and a beam of parallel rays.

Figure 10.11 The zone plate is a beam-forming device.

▶ **Example 10.3** The radial distances from the center of the central open zone to the inner edges of the next two open zones of a microwave zone plate are 14.1 and 20.0 cm. What is the focal length of the zone plate with 1.0 cm microwaves?

The diagonal distance from the first open zone to the focal point on the axis is $\sqrt{(f^2 + 14.1^2)}$. This distance is one wavelength longer than the distance along the axis from the centre of the zone plate to the focal point. Therefore, $\sqrt{(f^2 + 14.1^2)} = f + 1$, from which $f = 99$ cm.

Using the 20.0 cm open zone, $\sqrt{(f^2 + 20.0^2)} = f + 2$, from which $f = 99$ cm.

The zone plate shown in figure 10.11 is about one meter square. Zone plates for use with visible light are orders of magnitude

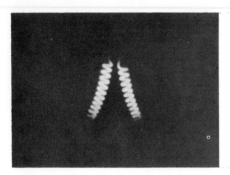

Figure 10.12 This photograph of a coiled filament was made with a zone plate.

smaller and are usually made by preparing an accurate black-and-white drawing on white paper and photographing the drawing with great reduction in size. Such an optical zone plate with only 12 zones produced the very respectable image of a coiled lamp filament shown in figure 10.12. The same microwave zone plate shown in figure 10.11 can be used as a focusing device for acoustic beams of frequency around 10 kHz and higher.

Zone plates, like most of the other focusing devices we have described, have focal lengths dependent on the wavelength of the radiation they are used with. They cannot be used effectively in broad-band situations, that is, with white light, wide-spectrum microwaves, or polytonal sound.

10.6 Mirrors

These drawbacks to the use of many of the beam-forming devices already considered have led to the widespread use of reflecting mirrors for focusing and for beam formation, in optics, microwave technology, and acoustics.

Consider the mirror surface sketched in figure 10.13. Suppose we want rays diverging leftwards from a point source s to be rendered parallel after reflection from the mirror to form a parallel beam towards the right. We wish to know the shape which the reflecting surface must have to accomplish this result. A spherical wave front expanding outward from s will be rendered plane after reflection if the distance from s to any point (x, y) on the mirror surface is equal to $s + x$. Therefore, $\sqrt{[(s-x)^2 + y^2]} = s + x$, from which

$$y^2 = 4sx. \tag{10.1}$$

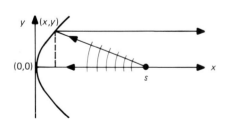

Figure 10.13 A parallel-ray beam is formed by reflection from a paraboloidal mirror.

This is the equation of a parabola centered at $(0, 0)$ with its axis coincident with the x axis and its focus at s. Since the figure represents a plane section through the mirror and its axis, the complete mirror surface must be a paraboloid of revolution about the x axis. Such a mirror will produce a beam of parallel rays originating from the point source s. Alternatively, it will bring to focus at s a beam of originally parallel rays coming from the right.

Although the image formed by a paraboloidal mirror is subject to some of the off-axis aberrations exhibited by images formed with lenses, it is free from chromatic aberration. The focal length s is independent of the wavelength of the radiation; the mirror can

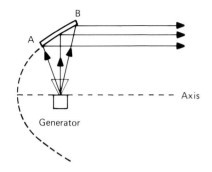

Figure 10.14 This off-axis reflector is called a horn antenna.

therefore be used in broad-band applications in optics, acoustics, and microwave technology.

Most large telescope objectives are mirrors. They have the advantage over lenses in that only one surface need be ground, shaped, and polished, whereas an achromatic compound lens has four surfaces. Also, in casting the mirror blank extreme care need not be taken to ensure uniformity of the index of refraction throughout. In microwave communication paraboloidal mirrors are used to produce the microwave beams on which are carried the speech, music, and digital data that constitute today's communication traffic.

A paraboloidal mirror used as a beam-forming device has a disadvantage in that the source generator and its housing, located on the axis of the mirror, block out some of the parallel rays reflected by the mirror. An adaptation of the paraboloid which overcomes this drawback is sketched in figure 10.14. The source generator, placed on the axis at the focus, irradiates an off-axis portion AB of the paraboloidal surface. This surface, with the mounting in which it is supported, is called a horn antenna. These structures can be seen ranged atop microwave towers all over the country.

10.7 Directional Antennas

When we think of beam formation at broadcasting radio frequencies, it must be clear that none of the devices so far considered would be at all effective. Now we are talking about wavelengths in the range from tens of meters to thousands of meters. Focusing structures of such dimensions are obviously out of the question. Nevertheless, in broadcasting antenna technology a directional quality can be imparted to the radiation emitted by making use of the phenomena of constructive and destructive interference. For example, figure 10.15 shows two vertical antenna towers seen from above. Suppose these towers to be located half a wavelength apart. If the antennas are powered by the same driving oscillator and are driven in the same phase, no radiation will be emitted along the line of the two units. This antenna pair radiates most strongly in the direction at right angles to the line between them. On the other hand, if the two antennas are driven 180° out of phase with each other, the radiated energy will concentrate in the direction of the line between them. A radio station located on a hill overlooking the city it is intended to serve can thus provide some selectivity in the direction of its radiation. Antenna arrays consisting of three or

Figure 10.15 Most of the energy is radiated sideways to the line joining these two antennas (top view).

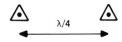

Figure 10.16 The antenna on the right is phased a quarter of a period behind the one on the left.

four units in line are even more directional and by proper phasing of the units the arrays can be made to radiate in 'side-fire' or 'end-fire' mode at will.

Unfortunately, these half-wave in-line arrays, directional as they are, radiate just as much energy in the backward direction as in the forward, or desired direction. To minimize this waste, other interference schemes can be invoked. Consider, for example, the antenna pair of figure 10.16, with its units a *quarter* of a wavelength apart and with the right-hand unit driven a *quarter* period behind the phase of the other. A little thought will reveal that this pair will radiate most strongly towards the right and not at all towards the left. However, it does emit quite strongly in both sideways directions as well. To suppress the sideways losses two such quadrature pairs are placed side by side half a wavelength apart, and the two pairs are driven in phase. The resulting radiation pattern is well concentrated towards the right, with no emission to the rear and only a small loss in the sideways directions (see figure 10.17).

10.8 Sound Beams and Images

You have probably never experienced what could properly be described as a true beam of sound. If you attempt to collimate a sound beam by projecting sound from a source through an open doorway, for example, you find you can hear the sound almost as well on the other side of the door whether you stand in the geometrically defined line-of-sight or well outside of it. This will be true even when you do the experiment under such conditions that reflection of the sound from walls or other objects can not affect the result. The reason, of course, is that sound of ordinary speech frequencies has wavelengths comparable with the size of the doorway and that, as discussed in §10.3, aperture diffraction is causing the sound to fan out on the other side of the doorway. If you were to repeat the experiment using sound in the frequency range of 10 kHz or higher, you would, indeed, be able to form a sound beam with a divergence of only a few degrees.

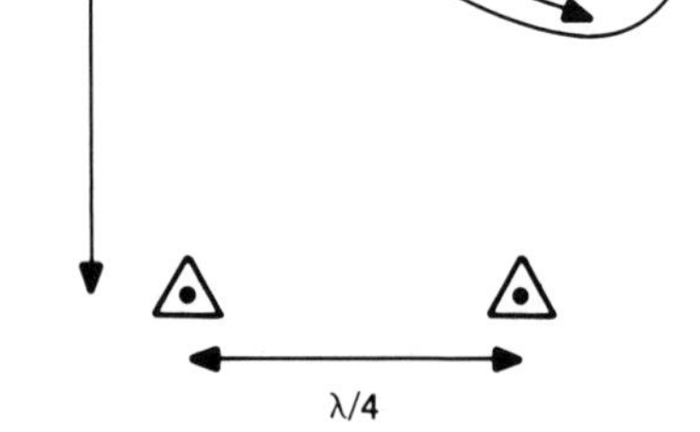

Just as a crudely collimated light beam can be produced by passing light from a source through a spherical flask of water, a crudely collimated sound beam will result from passing sound from a source through a spherical rubber balloon filled with gas, CO_2 for example, having a lower sound velocity than air. A better collimated beam can be obtained by making the refracting gas chamber more lens-like. Here, too, diffraction causes a divergence of the resulting sound beam, however well collimated geometri-

Figure 10.17 This four-unit array radiates mostly to the right.

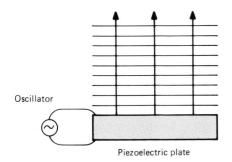

Oscillator

Piezoelectric plate

Figure 10.18 A parallel-ray beam of ultrasound is produced by a quartz-plate generator.

cally. To be effective at ordinary speech frequencies such a lens would have to be several meters in diameter.

Because of their high frequencies, however, bird sounds can be focused after a fashion by small paraboloidal reflectors. Equipment currently in favor for recording bird sounds includes a pick-up microphone at the focus of a portable 'dish' reflector about half a meter in diameter.

At very high frequencies, the focusing of sound beams with lenses and mirrors is standard laboratory practice. When ultrasound is produced by the vibration of a flat piezoelectric plate driven by an electrical oscillator (figure 10.18), the entire surface of the plate vibrates in the same phase, and the resulting beam of ultrasound is fairly well collimated as it is produced. The resulting sound beam can be focused with lenses or mirrors.

In medical applications ultrasound is employed to examine internal organs of the body. It is safer than x-rays and gives better contrast in delineating the softer tissues. Ultrasound in liquid propagation media can be produced with frequencies in the megahertz range and wavelengths comparable with the wavelengths of visible light. An acoustic microscope which 'sees' by ultrasound has been devised† which resolves detail in unstained blood corpuscles with better contrast than can be had with a light microscope.

10.9 Particle Beams

Today's technology embraces not only beams of radiation but also beams of particles. In a giant particle accelerator ions, electrons, or atomic nuclei are accelerated to high speeds in atom-smashing experiments. In every television picture tube a focused beam of electrons impinges on a fluorescent screen to produce the picture you see.

In molecular-beam experiments with particles which have no net electric charge, about the only way to produce a beam of particles moving in parallel paths is by geometrical collimation with apertures. In such an arrangement (figure 10.19), the first aperture acts as a source, and the second acts as a selector to exclude from the resulting beam all molecules whose paths are not oriented in the desired direction. The space between the two apertures is continually pumped out to ensure that the molecules that didn't pass through the second aperture don't build up a high

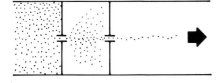

Figure 10.19 A molecular beam is geometrically collimated by two apertures.

† Wade G (1975) Acoustic imaging with holography and lenses *IEEE Trans. Sonics and Ultrasonics* **SU22** (6) 385

concentration there. The result is a beam of molecules whose paths are nearly parallel to each other.

Such geometrically collimated beams suffer the same disadvantage as geometrically collimated radiation beams: the better the collimation, the smaller the apertures and the lower the flux have to be. With charged particles, however, this disadvantage can be overcome by the use of larger apertures and electrostatic or magnetic focusing of the resulting divergent beams. Figure 10.20 shows the operation of an electrostatic lens. A divergent beam of electrons emitted from a source cathode S passes through circular holes in two metal plates in succession. The first plate is charged positively with respect to the cathode in order to draw electrons in the direction of the holes. The second plate is charged highly positively with respect to the first. The arrows between the plates show the configuration of the electric field, which has curved lines of force extending into the space between the holes. An electron moving straight along the axis of the holes will pass through them without any deflection. An electron moving along a divergent path entering the first hole will be deflected toward the axis by the radial component of the field in region A. It will be deflected away from the axis by the field in region B. However, the electron is going faster in region B, and the away-from-the-axis deflection will be less than the toward-the-axis deflection in region A. The result is a net bending of the electron path back towards the axis, with which it finally converges in the bright focal spot on the screen, along with the paths of all the other electrons negotiating the holes.

Even though we are dealing with particles rather than waves, every particle optical system is subject to aperture diffraction, just as wave optical systems are. All particles are associated with matter waves (de Broglie waves) having wavelength $\lambda = h/p$, where h is Planck's constant and p is the momentum of the particle. These matter waves are diffracted by any aperture which defines the beam, with the result that there is always some

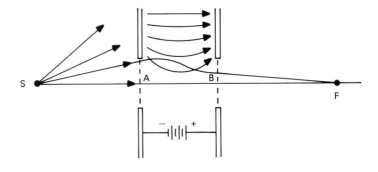

Figure 10.20 Coaxial holes in charged metal plates form an electron lens.

divergence even in the best-collimated particle beam and some fuzziness in the best-focused particle-formed image.

Wave properties of the electron are exploited by the electron microscope. This instrument forms images with particle rays rather than light waves. In any microscope it is impossible to produce a good image of an object which is comparable with or smaller than the wavelength of the radiation by which it is illuminated. Electrons accelerated by high voltages have wavelengths hundreds of times smaller than visible-light wavelengths, so the possibility exists of electron microscopes which will resolve image detail hundreds of times finer than can be resolved by a light microscope (figures 10.21 and 10.22). Electron microscopes

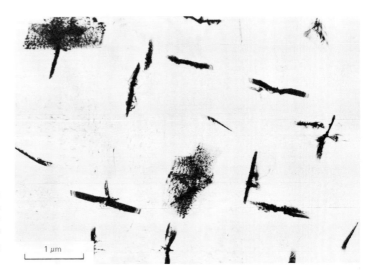

Figure 10.21 An electron micrograph showing precipitants of vanadium carbide in a vanadium sample containing 0.9 atomic percent carbon. The sample has been radiation cooled and aged for 120 minutes at 350 °C. (Courtesy of Professor Peter A Thrower, Pennsylvania State University.)

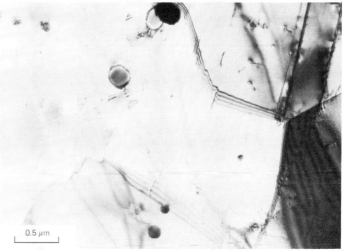

Figure 10.22 Structure of a thin stainless-steel foil. The parallel fringes are produced by grain boundaries passing obliquely through the foil. The short wiggles are dislocations which also pass through the foil—the ends correspond to points of intersection with the top and bottom of the foil. Round black regions are carbide precipitates. (Courtesy of Professor Peter A Thrower, Pennsylvania State University.)

have been built which have useful magnifications of 100 000, whereas the useful magnification of a light microscope is around 1000.

In magnetic focusing, a magnetic field is imposed on the electron beam over its entire length, with the field direction along the axis of the beam. An electron which leaves the source at an angle to the axis has a radial component of its velocity at right angles to the field. An electron moving at right angles to a magnetic field follows a circular path, which, combined with the along-the-axis component of the electron's motion, causes it to trace a spiral trajectory. This spiral brings the electron back to the axis at a distance from the source such that the spiral has made a complete revolution. Figure 10.23 shows a perspective sketch of several such spiral paths in a magnetically-focused beam.

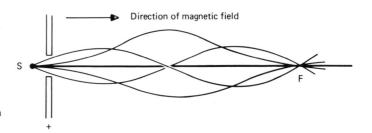

Figure 10.23 The electron paths in a longitudinal magnetic field are spirals.

10.10 Summary

What common threads can we find in the content of this chapter on beam formation? For beams of wave radiation the common thread is the production of wave fronts which have essentially zero curvature. This can be accomplished by (i) generating the waves with flat wave fronts to begin with, (ii) taking a bundle of radially diverging rays and transmitting them through a structure which either selectively retards the central rays or advances the peripheral rays, or (iii) reflecting a bundle of radially divergent rays from a mirror of suitable shape. For particle beams other devices must be resorted to, from simple geometrical collimation to electrostatic and magnetic focusing.

For both wave and particle beams, however well collimated from the geometrical point of view, there is a limit to the achievable parallelism of the rays or particle paths set by the ratio of the wavelength to the diameter of the defining aperture.

► **Exercises**

10.1 If the searchlight beam in figure 10.2 does not get substantially dimmer the farther you travel along it, isn't there a violation of the inverse square law? Explain.

10.2 Develop reasons for believing that the lens formula,

$$\frac{l_0}{l_i} = \frac{d_0}{d_i},$$

should or should not be valid for zone-plate optics. Here l refers to the length of an object or image; d refers to its distance away from the lens.

10.3 A positive lens reduces the curvature of wave fronts passing through it by retarding the central rays more than the peripheral rays. With a zone plate, however, it is clear that peripheral rays, following longer optical paths, are retarded more than the central rays, yet the result is an output wave front of reduced curvature. How can this be?

10.4 All the lens defects discussed in §10.3 for glass lenses have counterparts for the electrostatic or magnetic lenses used in an electron microscope. (a) Explain the meaning of 'chromatic aberration' for an electron lens. (b) In a light microscope a compound objective is used to minimize spherical aberration, yet a simple lens is used for an electron microscope. Why is spherical aberration usually no problem in an electron microscope? HINT: The objective may have a focal length of about 2 mm and an aperture of 2×10^{-3} cm.

10.5 Would it be possible to make a microscope in which a proton beam illuminates the specimen? Suggest possible advantages and disadvantages that a proton microscope would have in comparison with an electron microscope.

10.6 A paraboloidal microwave reflector has a diameter of 5.0 m. What is the approximate spreading angle of the best collimated beam it can produce with 6.0 cm microwaves?

10.7 The huge Arecibo radio telescope has an aperture diameter of 300 m. What is the smallest angular separation of radio sources in the sky that can just be resolved using the 21 cm radiation from hydrogen?

10.8 A plano–convex lens uncorrected for chromatic aberration is made of dense flint glass having indices of refraction of 1.643 and 1.674 for red and violet light respectively. The diameter of the lens is 5.0 cm and the radius of curvature is 10.0 cm. What is the diameter of the circle of least confusion due to chromatic aberration when the lens is used with white light?

10.9 Show that the zone plate of Example 10.3 has, in addition to the focal length of 99 cm, other focal lengths of 49 cm, 32 cm, ...

10.10   The plates of a microwave louvre lens have a separation such that the secondary wave trains A and B moving with velocity c (figure 10.9) intersect each other at 90° and the plate surface at 45°. Show that the phase velocity of the waves in the lens is 1.414c and that the index of refraction of the lens is 0.707.

10.11   Show that if an electron has been accelerated through a potential difference V (in volts) in an electron microscope, its de Broglie wavelength λ (in angstroms) is given by,

$$\lambda = \frac{12.3}{\sqrt{V}} \text{ Å}.$$

Filters and Frequency Shaping

11

. . . the universe is not only queerer than we suppose but it is queerer than we can suppose.

J B S Haldane

11.1 Examples of Filtering

In this chapter we shall treat yet another aspect of wave physics, dealing with the shaping of the frequency spectrum of a wave source or transmission medium by means of devices called filters.

Examples of filter action are numerous in nature and still more so in man-made devices. For example, the setting sun appears red although it is actually white; the crackling crash of a nearby thunderbolt is softened to a rolling rumble at distances of a couple of miles or more; high-quality radio receivers are provided with controls whereby the listener can accentuate bass or treble as he desires; the driver of an automobile speeding along a rough track is isolated from the road vibration by a spring-and-mass suspension system; and your voice sounds different when you shout into a bucket instead of directly into the air.

In preceding chapters, for the sake of simplicity, we have dealt with wave sources which generate waves of one or two discrete frequencies, usually of sinusoidal shape. However, many wave sources emit waves of several frequencies simultaneously in a continuous range of frequencies called a spectrum. A symphony orchestra, a babbling brook, a waterfall, and ordinary street-noise sources afford examples in the acoustic realm. In optics there is the 'white' light from an incandescent source and the thermal radiation from a hot soldering iron. In electricity there is the spectrum of signals coming over a carrier telephone line. The vibrations of an earthquake sequence or of an automobile being driven over a road with random bumps afford examples of continuous spectra in mechanics.

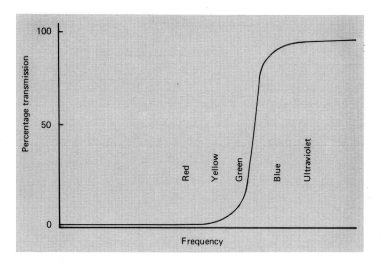

Figure 11.1 The transmission of a color filter varies with frequency.

In many instances in physics and engineering it is required to take the output from some broad-spectrum source and reshape it, accentuating certain frequencies, diminishing others, perhaps completely supressing still others. This spectral reshaping is accomplished with devices called filters. A practical example of filtering is the production of a beam of blue light by sending a beam of white light from a filament source through a blue filter which absorbs the green, yellow, and red portions of the spectrum (figure 11.1). Frequency filters are found in electrical, acoustical, and mechanical systems as well. Their operation and uses are the subject of the following sections.

11.2 Electrical Filters

Every AC-operated television and radio receiver has one or more filters. The 60 Hz (USA) or 50 Hz (UK) AC supply power is fed into a rectifier which converts the alternating current into a unidirectional but pulsating current output at 120 Hz (USA) or 100 Hz (UK). The rectifier output can be regarded as the resultant of a DC component, given by the broken line in figure 11.2(right), upon which is superposed a 120 Hz (100 Hz) ripple, along with a string of higher harmonics of 120 Hz (100 Hz) due to the non-sinusoidal shape of the rectifier output wave. Before this output can be used to operate the tubes or transistors of the receiver, the pulsating flow must be smoothed out into a steady direct current. This smoothing is accomplished by transmitting the rectifier output through an electrical filter which allows the passage of low frequencies and direct current (zero frequency) but blocks the

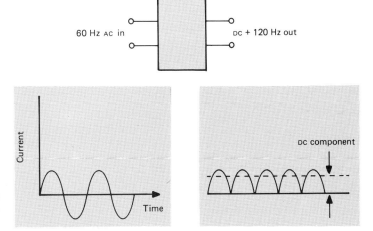

Figure 11.2 The output of a 60 Hz rectifier contains a large-amplitude component at 120 Hz.

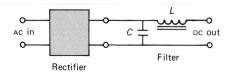

Figure 11.3 Some smoothing is afforded by this one-stage filter.

transmission of the alternating current components at frequencies of 120 Hz (100 Hz) and higher.

How is such an electrical filter constructed? You might make a first attempt by simply including a large self-inductance coil in the rectifier output circuit. An inductance presents low impedance to alternating currents at low frequencies and high impedance to alternating currents at high frequencies. Such a coil ideally would transmit the DC component of the rectifier output readily, but would reduce considerably any AC ripple. However, as a practical matter, the inductance coil required to reduce the 120 Hz (100 Hz) ripple to a level below the annoyance threshold would be much too bulky and heavy. Also, the wire winding of such an inductor would have prohibitively large DC resistance as well.

How can we improve the performance of the inductance-coil filter using circuit elements of reasonable size, weight, and cost? One improvement is suggested by the observation that a capacitor has low impedance at high frequencies and high impedance at low frequencies. Thus by placing a large capacitor *across* the rectifier output while placing the inductor *in series* with it, one might partially short-circuit the 120 Hz (100 Hz) ripple before it reaches the inductor. A one-stage *L–C* filter of this construction (figure 11.3) is more effective in reducing the AC ripple than an inductor of many times the inductance value alone.

Still further reduction of the unwanted ripple may be secured by cascading two or three stages of the simple *L–C* unit, as shown in figure 11.4. In this way the residual AC hum in a good radio receiver can be reduced to less than one millionth of the useful sound power the receiver is intended to process. The inductors and capacitors need not be burdensomely large, massive, or costly.

It is not the purpose of this chapter to develop the mathematics of filter theory, but rather to show that filtering is an across-the-board phenomenon appearing analogously in electricity, acoustics, optics, mechanics—wherever waves and oscillators are found. However, one particular outcome of a mathematical analysis of a multi-stage filter such as that of figure 11.4 is that there is very little reduction in the transmission for frequencies below a certain critical cut-off frequency f_c, while for frequencies above the cut-off frequency the transmission falls off very

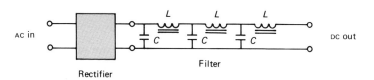

Figure 11.4 This multi-stage filter eliminates AC hum.

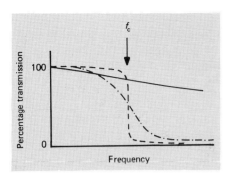

Figure 11.5 As the number of stages is increased, the frequency cut-off is sharper. Full curve, inductance alone; broken curve multi-stage; chain curve, one-stage L–C.

sharply. These comparisons are shown in figure 11.5. The sharpness of the cut-off increases with the number of stages. Because this filter allows the unimpeded transmission of DC and low-frequency AC, it is called a low-pass filter. The critical frequency for cut-off is given by $f_c = 1/\pi\sqrt{(LC)}$. Thus, for a 120 Hz (100 Hz) power pack ripple filter having several stages of $L = 10$ H and $C = 10$ μF, the cut-off frequency would be around 30 Hz, well below the frequency of the 120 Hz (100 Hz) ripple it is desired to block.

The cut-off frequency can be fixed by any combination of L and C which gives $LC = 1/\pi^2 f_c^2$. In choosing values of L and C to fulfill this requirement, it is preferable to choose smaller values for L and larger values for C because it is easier to make large capacitances than to make large inductances; also a large inductance of manageable size would unavoidably entail high DC resistance in the wire coil.

By reversing the positions of the inductors and capacitors one can obtain a filter which transmits, with very little attenuation, all frequencies *above* a certain critical frequency, while blocking the transmission of all frequencies lower than f_c. Such a filter is called a high-pass filter. More complicated repeated networks employing inductors and capacitors permit the transmission of a selected band of frequencies while blocking the transmission of frequencies both above and below the edges of the transmitted band. These are called band-pass filters. Finally, still other types of L–C assemblages yield band-stop filters which transmit all frequencies except those within a selected band.

These more esoteric filters have practical uses. In the early days of the wax-disc phonograph recording, the scratch of the needle in the groove proved very objectionable, often masking the softer passages of the recorded music. It was found that most of the scratch was centered on a relatively narrow range of frequencies towards the upper end of the audio spectrum. This range could be cut out with a band-stop filter in the phonograph amplifier without intolerably degrading the overall quality of the music.

When attempting to detect a very faint radio signal in the presence of other unwanted signals covering a wide spectral range, one can build into the receiver a band-pass filter which permits the reception of the wanted signal at its own particular frequency while rejecting the noise at higher and lower frequencies. In fact, the intermediate-frequency stages of a superheterodyne receiver perform such a filtering action, to reject radio signals other than the particular one to which the receiver is tuned.

Band-pass filters are also used extensively in carrier telephony to separate the voice channels carried together on a single microwave beam or coaxial cable.

An interesting application of filter theory is the inductive loading of telephone lines. A telephone line consists electrically of more than just two parallel conductors. Each conductor has series resistance and inductance, and between the two conductors there is shunt capacitance and insulation leakage conductance. The effect of this electrical constitution is to make a transmission line act somewhat like a low-pass filter. Because of the presence of the series resistance and shunt capacitance, the cut-off frequency is not sharply defined, and the line exhibits a smeared-out frequency response with progressively more severe attenuation at higher frequencies. This kind of response results in the distortion of the transmitted signals, which can be especially damaging if the line is to be used for carrier transmission with frequencies up to several tens of kilohertz. By deliberately inserting additional series inductance into the conductors, telephone engineers can produce a line with a sharper cut-off frequency and a flatter response characteristic in the range of frequencies below the cut-off. The result is more distortion-free transmission over the useful frequency range. The additional inductance is provided by splicing inductance coils in series with the conductors every few thousand feet. These are called loading coils, and the resulting line is said to be loaded.

11.3 Mechanical Filters

A frequently encountered problem in mechanics is that of isolating a delicate piece of apparatus from building vibrations. Remembering that springs and masses are the mechanical entities analogous to capacitors and inductors (see table 6.1), we would expect that a one-stage low-pass mechanical vibration filter could be made by hanging a mass from the ceiling on a spring and using the mass as a support for our apparatus. In designing such a support we would choose spring and mass parameters guaranteeing that the natural frequency of the system would be well below the lowest building vibration frequency from which we wished to protect the apparatus.

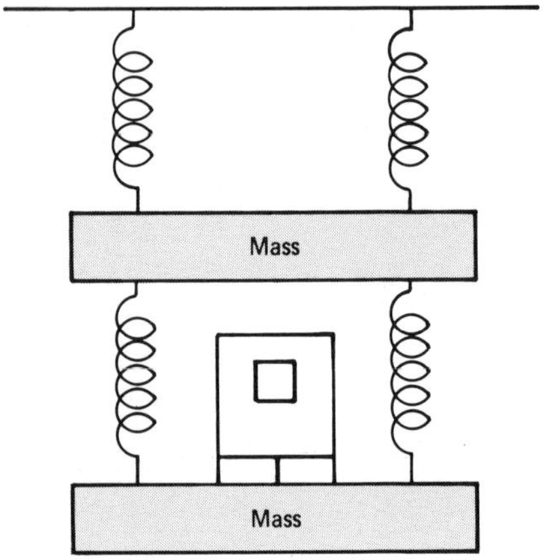

Such a system is, of course, an oscillator, as we have described in Chapter 6. The response of the mass to the forced vibrations applied at the upper end of the spring would result in a small, but finite, amount of vibration feeding through. To attenuate this residual vibration still further, a second stage of filtering should

Figure 11.6 This mechanical filter produces a vibration-free support.

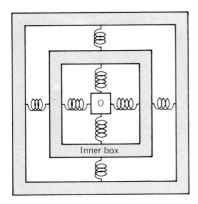

Figure 11.7 The packaged object is protected from high-frequency vibrations.

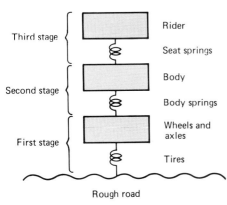

Figure 11.8 The rider is protected by a three-stage filter.

be used. Indeed, in practice, vibration-free galvanometer mountings customarily are made with two stages, as shown in figure 11.6.

Sometimes when a delicate object is to be packaged for shipment, it is floated with foam padding in an inner box having considerable mass. This inner box, in turn, is packed into the outer box with more padding. The complete package thus takes the form of a two-stage low-pass filter, shielding the object from the high-frequency components of the mechanical shocks that would do most damage (see figure 11.7).

The spring-and-mass suspension of the rider in an automobile affords a three-stage low-pass mechanical filtering of road vibrations. In figure 11.8 the tires and running gear constitute the spring and mass elements of the first stage of the filter. The body springs and car body constitute the second stage. Finally, the seat springs and the rider himself form the third stage. Small wonder you feel little vibration when you drive your car over a cobbled road. This system is a low-pass filter, with a cut-off frequency below the frequencies of vibration of a rough road surface.

In hydraulics it is sometimes desired to smooth out the pulsating flow of water coming out of a reciprocating pump. This smoothing is accomplished with a low-pass hydraulic filter analogous to the low-pass *L–C* electrical filter. The hydraulic filter is made with one or more stages each consisting of a length of pipe with a side chamber partly filled with air. The mass of the water in the main tube corresponds to the electrical inertia of the inductor of the electrical filter, while the springiness of the air in the side chambers corresponds to the compliance of the electrical capacitor. Figure 11.9 shows a two-stage filter of this construction. Its critical frequency is designed to be lower than the cycle frequency of the pump.

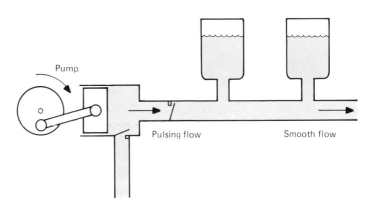

Figure 11.9 Two stages of low-pass hydraulic filters smooth the water flow from the pump.

(a)

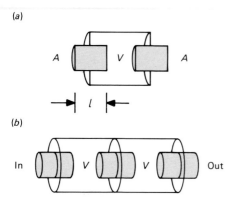

(b)

In   V   V   Out

Figure 11.10 An automobile muffler usually contains one or more stages of a low-pass filter as shown.

11.4 Acoustic Filters

The silencing of the exhaust noise from internal combustion engines affords perhaps the most widespread use of acoustical low-pass filters. A single stage of such a filter consists of an input pipe of cross sectional area A leading into a volume V, with an output pipe of the same cross section leading out of that volume (see figure 11.10a). A first-class automobile muffler may have two or three stages of such filtering contained in the muffler housing (figure 11.10b). The low-pass cut-off frequency of a filter made of a large number of such stages is given by

$$f_c = \frac{1}{\pi} \sqrt{\frac{v^2 A}{lV}},$$

where v is the speed of sound and l is the length of the connecting tube sections. Compare the geometry of this filter with that of the hydraulic filter of the last section. Can you identify the corresponding elements?

11.5 Energy-absorbing Filters

The filters we have been considering are frequency selective because outside the band of readily transmitted frequencies they operate by *reflecting* energy back towards the source. They do not absorb the unwanted frequencies, as they are constructed with non-dissipative elements. In the optical domain, on the other hand, most filters operate by selectively absorbing the unwanted spectral regions and transmitting the rest. Thus camera filters, eyeshades, tinted glasses, and the dyes used in color printing are energy-absorbing filters. For devices such as these the filtering action depends on the quantum absorption processes of individual atoms, ions, and molecules. These filters can hardly be compared, element by element, with the electrical, mechanical, and acoustical filters we have already considered, because quantum processes are so different in nature from those of wave physics. In Chapter 9, we encountered optical frequency-selecting behavior that does depend on the wave properties of radiation. Interferometry, beam formation, and filter theory have many features in common.

Before we leave the subject of energy-absorbing filters, we would like to describe a particular filter which does absorb energy, the so-called resistance–capacitance (R–C) filter. We shall here treat the electrical version of this filter which is often used for smoothing out the ripple from power-supply rectifiers when the

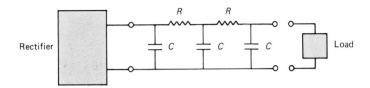

Figure 11.11 This filter absorbs energy rather than reflecting it.

current drain is small. In the circuit of figure 11.11, at each cycle of the rectifier output, the input capacitor of the filter is charged up to the peak voltage of the rectifier. During the rest of the cycle the charge on the capacitor leaks off into the following stages of the filter and, eventually, into the load.

The transmission against frequency characteristic of this filter does not exhibit a sharp cut-off, the transmission merely becomes smaller and smaller at increasing frequencies. As more current is drawn through this filter, its attenuation of the 120 Hz (100 Hz) ripple becomes less effective unless the capacitance values are correspondingly increased.

11.6 Filtering by Periodic Media

You may have noticed, while thinking about the various filters we have been describing, that they are all specially tailored sections of wave-propagation media and that they all consist of successive tandem repetitions of some simple structural unit. You might have been led to wonder whether any wave medium which exhibits a periodically repeated structure would have some frequency-shaping effect on a spectrum of waves launched into or onto it. The answer is yes, it would. The magnitude of the effect will depend on the frequencies contained in the particular spectrum and on the precise nature of the repeating unit in the structure of the medium.

Some insight into the theory of the repeated-unit filter can be developed with reference to the propagation of waves on a vibrating string loaded with extra masses at equal repeated intervals of position. To avoid the complication of reflections from the far end of the string, let us suppose that it is infinitely long. Any individual wave or pulse traveling along the string will be partly reflected at each successive mass point and will suffer an overall exponential decay as it travels farther along the string. The energy of the wave is reflected back, bit by bit, towards the generator.

If the waves are launched in a continuous train, the same process of reflection will occur with each wave. The waves partly reflected at the mass points will be partly reflected again at

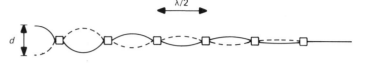

Figure 11.12 The waves die out rapidly when $d = \frac{1}{2}\lambda$

adjacent mass points, and so on. You might expect the wave pattern on the string to become very complicated, with partly reflected waves coming and going in both directions. However, if the wavelength of the wave bears no simple ratio to the distance between the mass points, the reflections traveling backwards towards the generator will, on average, superpose *destructively*, and no net reflection of energy will occur. The original waves will appear to travel down the string with undiminished amplitude. Even though reflection and re-reflection are taking place all the time, the partly reflected waves traveling backwards towards the generator will be phased so as to prevent any strong back-reflection from building up.

Only if the waves have such frequency that the distance between mass points is exactly or nearly equal to half a wavelength does any substantial reflection occur (figure 11.12). In this case all the partially reflected waves superpose *constructively* in the backward direction to produce a strong reflected component which takes all the energy from the original waves. The original waves disappear after penetrating only a few spacings into the medium. The transmission against frequency characteristic of the loaded string therefore exhibits a deep drop in the neighborhood of this critical frequency. Figure 11.13 shows

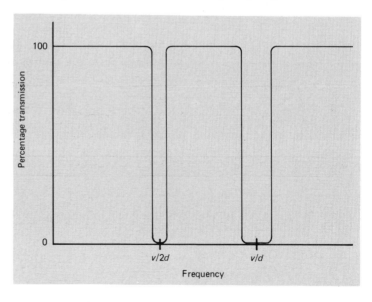

Figure 11.13 The mass-loaded string is a band-pass filter.

such a characteristic, which you will recognize as that of a band-stop filter. The mid-band frequency of this stop gap is given by the condition $d = \frac{1}{2}\lambda$, where d is the distance between adjacent mass points. Since $\lambda = v/f$, where v is the speed of the waves on the string, this condition can be written $f = \frac{1}{2}(v/d)$.

In Chapter 8 we developed an expression showing that a string of length d resonates at the frequency for which $d = \frac{1}{2}\lambda$. Thus, for our mass-loaded string, we see that the stop band occurs at the frequency for which each segment of the string between mass points becomes resonant.

A particularly interesting example of multi-stage filtering by a periodic wave medium is the structure of the energy-level spectrum of the valence electrons in solid crystalline materials. We are here concerned with the motions of the electrons as they move about in the interior of a solid. Each electron behaves as though it were representable by an electron wave propagating through the crystal ion lattice, where the ions present a periodic potential through which the electrons move. We thus have a case of waves traveling through a periodic medium. The ion lattice acts like a multi-stage band-stop filter permitting the motion of electrons having certain wavelengths and disallowing the motion of electrons with wavelengths falling within the stop band. Since the wavelength of an electron (de Broglie) wave is related to the energy of the electron, it follows that electrons having energies falling in certain ranges may move freely within the crystal while electrons in certain other energy ranges are forbidden. Figure 11.14 shows this band-stop structure of the energy spectrum of the

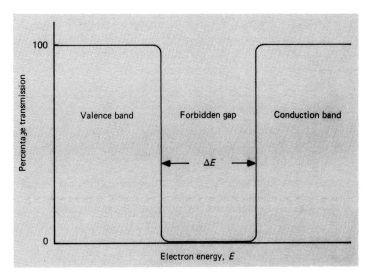

Figure 11.14 Electrons are not allowed to move with energies in the forbidden gap.

valence electrons in crystalline solids. For many solids this spectrum exhibits a vacant region, called the 'forbidden gap', depicting a range of energies which the electrons are not allowed to populate.

This forbidden energy gap, or stop band in filter parlance, has important implications in the model that we shall develop in Chapter 14 for the electrical conduction mechanism in solids. The width of the gap ΔE and the electron densities in the bordering allowed pass bands determine whether a solid is a conductor, a semiconductor, or an insulator.

► **Exercises**

11.1   How do you react to the suggestion that when a prism spectroscope is used to isolate a particular spectral region, it is really being used as a band-pass filter?

11.2   The vascular system of the body consists of the heart and a low-pass network of arteries in which the distributed elasticity of the vessels' walls takes the place of the discrete air chambers of the filter of figure 11.9. How is the critical cut-off frequency of this system related to the heartbeat frequency?

11.3   If you try to run your car in high gear at too low a speed, the car will buck at each firing of the cylinders of the engine. This bucking does not occur at higher speeds. Why?

11.4   In a steam-driven train the locomotive accelerates by a series of jerks occasioned by the successive thrusts of the piston rods. The passengers in the cars, however, feel only a steady acceleration. Why?

11.5   When driving a boat into waves whose wavelength is comparable with the length of the boat, one encounters very uncomfortable pitching. How might you proceed to reduce the pitching?

11.6   The earthquake vibrations which do most damage to buildings are concentrated in the frequency range below about 3 Hz. How would you design an earthquake protection system for a five-storey building with a mass of 1500 tonne? Consider both vertical and horizontal vibrations of amplitude up to several centimeters.

11.7   A shaft rotates continuously but with much superposed angular oscillation due to the meshing of the gear teeth which drive it. Design a mechanical filter which will damp out the irregularities in the drive.

11.8   A steady flow of high-pressure air is required for a sprayer. The pressure is provided by a reciprocating piston pump at 5 strokes per second. Design a smoothing filter for the air stream.

11.9   From the broad-spectrum output of a hi-fi amplifier it is desired to filter out the signals below about 200 Hz and send them to a bass-range woofer speaker. The signals above about 2 kHz are to be filtered out and sent to a tweeter speaker. The rest goes to a mid-range speaker. Design a filter network to accomplish this separation.

12 Transformers and Impedance Matching

. . . arrange better impedance-match between science and people. That's what relevance is—educational impedance matching.

Eric M Rogers

12.1 Introduction

Your first encounter with the devices called transformers un-
doubtedly came as part of the electrical portion of your course in
introductory physics. You learned that transformers are used to
increase or decrease the voltage supplied by an AC source to make
it satisfy the operating voltage requirement of some device
consuming AC power. However, before you met the name
transformer, you encountered the idea of transformers much
earlier in your experience. Perhaps this introduction came when
you used a crowbar to move a heavy object, or when you were
learning how to ride a bicycle and master the technique of
changing gears, or perhaps still earlier when you discovered that
you could swim much faster with a pair of rubber flippers on your
feet. If you own a camera or a pair of binoculars with coated
lenses, you have been using transformers for longer than you
realize.

What is the connection between these seemingly quite different
devices? Their family designation—transformers—suggests that
they are used to change something, and indeed they are. In
different areas of physics and engineering there are transformers
which change the magnitude of voltages, currents, forces, velocities, pressures, and fields, in order to achieve useful ends.
Transformers have two principal applications. They are used for
their step-up and step-down capabilities. They are also used as
impedance matching and mismatching devices to maximize or
otherwise control the transfer of power from a source to a power-
consuming load. Analytically, both these functions are aspects of
the same behavior. However, we shall consider them separately.

12.2 Step Up and Step Down

In your earliest formal acquaintance with electrical transformers
you learned that you could operate a 6 V lamp from a 110 V
(USA) or 220 V (UK) AC supply by inserting a suitable voltage
step-down transformer into the circuit. Perhaps you learned also
that to accomplish this step-down function the primary and
secondary windings of the transformer had a turn ratio equal to
supply voltage/lamp voltage. Any attempt to operate the lamp
directly from the supply would result in such high current as to
burn out the lamp in a single flash.

The efficiency of a transformer is never 100 %, but it may be
high, say 98 %. In an ideal (100 % efficient) transformer the
voltage transformation is always accompanied by a current

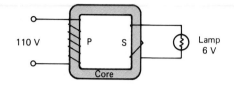

Figure 12.1 A step-down transformer operating a 6 V lamp from a 110 V supply. P, primary; S, secondary.

transformation of the inverse ratio. The step down of voltage in the case illustrated (figure 12.1) is accompanied by a step up in current by the factor 110/6. That is, if at 6 V the lamp current is 1 A, the supply current in the primary of the transformer is only 6/110 A. The power delivered to the transformer primary circuit is therefore the same as the power transmitted to the lamp, with some allowance being made for a small amount of power to be lost in the process.

The ideal transformer, of course, does not exist in actual circuits. There is always some power lost to eddy currents in the transformer core and to Joule heating in the resistances of the primary and secondary windings. However, modern transformers can be made with internal losses as small as 2 % of the power they transmit.

Transformers are used extensively in electrical technology. Electric power is most conveniently produced by moving-coil dynamos at a few hundred or a few thousand volts. For household consumption, however, safety considerations call for much lower voltages in house circuits, usually 110 V in the USA, 220 V in the UK. On the other hand when electric power is to be sent some distance on a transmission line, it is customary to transmit at high voltage and low current in order to reduce the $I^2 R$ Joule-heating losses in the wire of the line. Transmission voltages of 55 000 or 110 000 V are common. Even higher voltages would be preferable, but a reasonable compromise must be made with the difficulty of providing insulation and preventing losses from the conductors. Losses may occur in arcing across the ceramic insulators or in the often visible corona, a discharge through air which has been ionized by the high field near the wires.

The versatility of AC power distribution is apparent in a doctor's office where the supply may be called upon to operate a $4\frac{1}{2}$ V pilot light on the incubator, provide 110 V illumination, deliver 220 V power to the sterilizer, and furnish 100 000 V for the x-ray machine. All these voltage steps up and steps down are accomplished with transformers.

You might remark at this point that if you wanted to operate a 6 V lamp from a 110 V or 220 V supply, you might do it by eliminating the transformer and simply placing a suitable dropping resistor in the lamp circuit (figure 12.2a). True, the lamp would light all right, but think of the power wasted in the resistor. By using the transformer, you draw from the source only the power required to operate the lamp itself, plus a small amount to make good the transformer loss.

The operation of electrical AC transformers makes use of

(a)

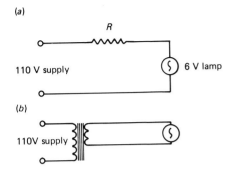

Figure 12.2 Which of these lamp circuits consumes less power?

Faraday's principle of induction which says that the voltage induced in the secondary winding is proportional to the rate of change of magnetic flux Φ, that is $d\Phi/dt$, produced by the alternating current in the primary winding. In DC circuits dI/dt and $d\Phi/dt$ are zero, so induction transformers are useless for direct current. Are there, then, no DC electrical transformers?

Yes, there are, but they are more complicated than AC transformers, and they are not suitable for handling power at kilowatt levels. One scheme involves interrupting the DC voltage to obtain a square-wave AC output, stepping up the voltage of this output with a conventional AC transformer, rectifying the result, and passing it through a smoothing filter. This scheme is used in the power converters for automobile radios operating from the 6 V or 12 V battery.

Another scheme involves a variable capacitor (figure 12.3) consisting of parallel rotor and stator plates shaped as sectors of a disk. When the plates are superposed, the capacitance is maximum and at this instant the capacitor is charged by a charging brush contacting the rotating sector. A quarter of a rotation later when the plates no longer overlap, the capacitance is a minimum. The capacitor holds the same charge it acquired at maximum capacitance; from $V = Q/C$ the voltage between the plates will now be many times higher. At this instant the capacitor is discharged by a take-off brush into the high-voltage storage circuit.

A third scheme can perhaps be described better after discussing its mechanical analog, the hydraulic ram (figure 12.4a). Water from a reservoir runs downhill through a sloping pipe. As soon as the speed of the water flow exceeds a certain critical value, a spring-loaded valve in the end of the pipe closes. The instantaneous pressure at the end of the pipe becomes very large due to the inertia of the moving water and the sudden shut-off of the

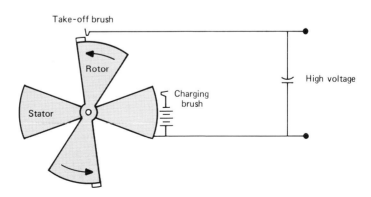

Figure 12.3 A rotating variable capacitance makes a DC transformer.

(*a*)   (*b*)

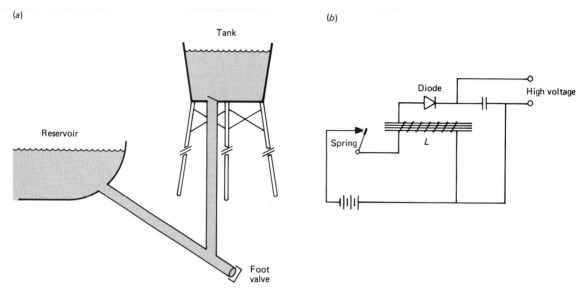

Figure 12.4 A hydraulic ram (*a*) has its analog in a circuit for raising a DC voltage (*b*).

flow. This upsurge of pressure forces some of the water up a side pipe into a storage tank at a higher level than that of the water surface in the reservoir. By this device some of the water attains higher gravitational potential than it had to begin with. Meanwhile the valve at the end of the pipe opens and the flow starts again.

The electrical analog of this device might be called an electrical ram. Remembering that inductance is the analog of inertia and that voltage is the analog of pressure, one can construct an electrical ram using a battery circuit which includes a relay with a highly inductive winding (figure 12.4*b*). The relay opens the circuit when the current exceeds a certain critical value. The breaking of the circuit causes an inductive surge of high voltage which forces charge into a capacitor from which its return is blocked by a diode. After a short interval the relay contacts close, and the cycle starts again with rising current in the battery circuit.

In mechanics the step-up and step-down functions are performed with the mechanical advantage machines you studied in your elementary physics course. The quantities transformed are force and velocity, which are the mechanical analogs of voltage and current. In the crank-and-axle machine (figure 12.5) used to raise a bundle of shingles to the roof, the force F exerted by the operator on the crank handle is stepped up by a factor R/r, but the speed with which the bundle rises is only the fraction r/R of the speed of the crank handle. By the use of this simple machine you can raise a much heavier load than you could manage

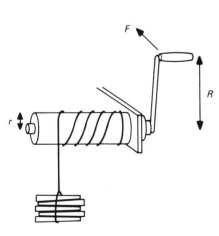

Figure 12.5 A crank and axle is a mechanical transformer.

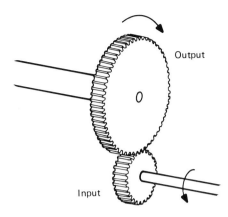

Figure 12.6 A gear train is a transformer for torque.

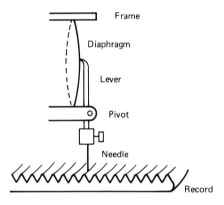

Figure 12.7 The lever in a mechanical phonograph pick-up is a transformer.

directly, hand-over-hand, but you pay for the step up of force by accepting a reciprocal step down in the speed with which the load is raised. For an ideal machine the power you expend is the same as the power consumed by the rising load. Actually the total work you do is the work done on the load plus a small increment to make up for frictional losses in the machine. Similar relationships can be worked out for the jackscrew (car jack), the inclined plane, the lever, and the gear train. Have you ever thought of these devices as part of the family of transformers that includes the electrical transformers as well?

In rotational mechanics the quantities analogous to voltage and current are torque and angular velocity. Torque transformation can be accomplished by gear trains or belt-and-pulley arrangements. In figure 12.6 the input shaft rotates at low torque and high angular velocity, while the output shaft is driven at high torque and low angular velocity. Such an arrangement is needed when an automobile is to climb a steep hill. If the motor were coupled directly to the drive shaft it could not furnish enough torque to propel the car up the hill. With the transmission in low gear, however, the motor torque is multiplied by a factor of two or three and the car keeps going, although at a slower speed than on a level road in high gear. Whatever increase in torque is provided by a torque transformer must be paid for by an equal factor of decrease in the angular velocity of the driven shaft.

We have described the mechanical transformers above as 'DC' devices employing steady forces, torques, and velocities. Such devices can also be employed in the transformation of quantities in time-varying or 'AC' motions. For example, in the old-fashioned mechanical phonograph pick-up, the needle is set-screwed into the end of a lever which is pivoted so that the 'AC' motion of the needle point is increased in magnitude by a factor of two or three at the point where it drives the sound-producing diaphragm (figure 12.7).

12.3 Impedance Control

To develop the subject of impedance matching and control we return to the first mention of impedance in Chapter 6, where we defined impedance as the ratio of voltage to current in an AC electrical circuit. Let's now continue farther with this idea. Every electrical circuit or device offers some hindrance to the flow of charge through it in response to the application of a voltage difference across its input terminals. In the same way a hose presents a hindrance to the flow of water in response to a pressure difference between its ends, as discussed in Chapter 1. The magnitude of this hindrance is called the impedance, Z.

Impedance is present in all physical situations, electrical, mechanical, acoustical, thermal, both AC and DC, where some effect ensues from a cause. It is present whether the result is steady with time (as in a DC electrical circuit with constant current) or whether the result is time-dependent, as in an AC circuit where the current fluctuates periodically with time. Impedance may be regarded as the ratio of cause to effect. In an electrical circuit the impedance is the ratio of the applied voltage to the resulting current. In a DC circuit, the impedance is numerically equal to the circuit resistance R, whereas in AC circuits the reactances of any capacitors and inductors must also be included in Z. In such cases the impedance is the vector sum of the resistance and the reactance:

$$Z = R + iX = R + i(2\pi f L - 1/2\pi f C).$$

One of the major concerns of engineering is the transfer of power from a source to a load: the transfer of power from a battery to the starting motor of an automobile, or from the engines of a steamship to the stream of backward-thrust water propelling the ship; transfer of acoustic power from a violin string to the surrounding air; transfer of power from a transmitting antenna to the surrounding space and its subsequent interception by receiving antennas near and far.

In many cases it is desired to maximize the power transferred, and the question is asked 'How can I get the most power out of this device—the maximum power it is capable of delivering?' Obviously you cannot operate an electric locomotive from a doorbell battery. Neither would you order a steam roller to crack a walnut. Nature seems to call for some kind of balance between the action taken and the implement you use to do it. The balance is set forth in the impedance theorem which says that the transfer of power from a source to a load is maximum when the impedance of the load is made equal to the internal impedance of the source.

▶ **Example 12.1** Prove the impedance theorem for a simple DC circuit consisting of a battery with EMF E and internal resistance r delivering power to an external load of resistance R (see figure 12.8a).

The current I in this circuit is $E/(R + r)$. The power delivered to the load is

$$P = I^2 R = \frac{E^2}{(R+r)^2} R.$$

(a)

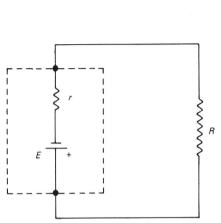

(b)

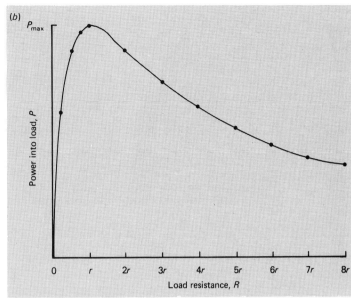

Figure 12.8 For the circuit shown in (a), power is a maximum when $R = r$ (b).

To find the resistance R at which this delivered power is maximum, differentiate the last expression with respect to R and set the result equal to zero, to obtain

$$0 = \frac{E^2}{(R+r)^2} - \frac{2E^2 R}{(R+r)^3},$$

$$= \frac{E^2(R+r-2R)}{(R+r)^3},$$

$$= \frac{E^2(r-R)}{(R+r)^3},$$

from which $R = r$ or $R = \infty$. The latter value, however, gives a minimum value of P. The graph (figure 12.8b) shows how the load power is dependent on the load resistance, with a maximum at $R = r$.

In the example above, one may make the observation that if the load has too high a resistance there isn't enough current to give good power transfer, whereas if the load has too low a resistance the resulting large current will pull down the battery-terminal voltage to the point where very little power is available for the load. When the impedance of the load is made equal to the internal impedance of the source the impedances are said to be matched, the power transfer is maximum, and the terminal voltage of the source is half its EMF.

12.4 More Impedance Matching

What has the foregoing to do with transformers? Transformers are employed to control the impedance matching between power sources and power-consuming loads in order to maximize or otherwise regulate the power transferred. In what follows we shall describe a number of instances of transformer action and point out the similarities and other features of interest.

In the design of musical instruments there is often a problem in transferring vibrations from the sound-producing element to the air. In the case of stringed instruments this transfer is very inefficient because the vibrating string, having a small cross section, 'cuts through' the air rather than seizing it and shaking it bodily. To facilitate the transfer of power from the string to the air, the instrument must be provided with an acoustic transformer: the piano with a sounding-board; bowed instruments with a hollow box or surface area comparable with the wavelength of the sound waves to be launched. The string transmits its vibrations to the transformer surface, which in turn launches them into the air. An impedance match is thus obtained, resulting in an enhanced transfer of power from the string into the air.

In most wind instruments the output of sound into the air is increased by flaring out the end of the tube to form a bell. The large flare typical of a brass instrument is an impedance-matching transformer between the tube and the outside air. Analysis is complicated by the fact that the matching unit is also an integral part of the primary vibrating unit. In the horn and most other brass instruments, the flare is approximately exponential. The cross sectional area increases at a constant ratio as we proceed by equal distances along the axis.

Consider a narrow cylindrical pipe with no flare. The discontinuity at the end is so great that most of the energy of a sound wave traveling along the pipe is reflected back, and only a small portion emerges. A flare improves the match between the pipe and the outer air; less energy is reflected from the discontinuity and more emerges. But if one produced a theoretically perfect match there could be no primary sound! Existence of a note of definite pitch depends on the reflection to set up standing waves in the pipe. As in the electrical analog, it is often easier to devise a tuned matching system which acts at only one frequency—obviously an undesirable limitation in a musical instrument.

A 'horn equation' evolved from work by Bernouilli, Euler, and Lagrange was part of the early blooming of the theory of partial differential equations which underlies much of physics. Their equation can be simplified by noting that as a wave travels into the enlarging part of a horn, the relatively high-pressure variations decrease because the sound energy is being spread over an ever wider front. If one extracts this part of the behavior of a wave in the horn from the mathematics of the horn equation, one is left with a much simpler equation, similar in form to Schrödinger's equation of quantum mechanics. The Schrödinger equation shows that a particle of energy E has associated with it a de Broglie wavelength λ that depends on the square root of the difference between the energy and the potential energy function V at any point in space:

$$\lambda = \frac{h}{\sqrt{(E - V)}}.$$

The simplified form of the horn equation leads to:

$$\lambda = \frac{v}{\sqrt{[f^2 - U(v/2)^2]}}.$$

At any point in the horn the acoustic wavelength λ depends on the square root of the difference between the square of the frequency f and a horn function U; v is the speed of sound. The horn function determines how much acoustic energy is reflected back to provide standing waves in the horn. The value of U at a given point depends on the interior radius r_i of the horn and the exterior radius r_e.

Regions where the horn function U is large can form a barrier to transmission of waves, hence reducing the escape of energy from the interior of the horn to the outside. Leaking of sound through the horn-function barrier is an analog to the leaking of a de Broglie wave and hence the penetration of particles through a nuclear potential-energy barrier in the radioactive decay of an atomic nucleus.

As one might expect from the discussion of horn flare, the shape of the mouthpiece is important in matching impedance at the entrance to the horn and in determining which resonance peaks are boosted.

In mechanics, impedance-matching transformers find wide application. The turbine of a steamship develops relatively low torque and must therefore run at high angular speed to produce enough power to drive the ship. The propeller, on the other hand, must have blades large enough to grip the water effectively and must expend its power at high torque and relatively low angular speed to avoid cavitation of the water. An impedance-matching gear train is interposed between the turbine and the propeller to permit the former to deliver its power effectively to the latter.

The muscles of a man's arms and legs can develop considerably more power than that needed to propel his body through water at 1 ms^{-1}. However, the size and shape of his hands and feet do not permit him to get a good enough purchase on the water to make use of this latent power effectively to propel himself. A pair of rubber frog flippers on his feet give him a better grip on the water and thus permit him to use more of the power his muscles are inherently capable of delivering. With the aid of such matching devices he can go much faster. Nature has implemented the law of impedance matching by evolving webbed feet for swimming birds.

In wave physics, too, impedance-matching transformers are often found. A wave of any kind launched onto a medium of any kind requires transfer of energy from the wave source to the medium. This transfer can be described with reference to the impedance of the medium. For an electrical transmission line, for example, the impedance is defined as the ratio of the AC input voltage to the resulting current in the line. For sound propagation the impedance of an air column in an acoustic tube would be the ratio of the driving force of the sound-producing diaphragm or piston to the resulting velocity of the layer of air adjacent to the vibrating surface. Here force F and velocity v are sinusoidally varying quantities. We refer to their RMS values, the square root of the mean (average) of F^2 or v^2 taken over one cycle. For a transverse mechanical wave whipped onto the end of a stretched clothesline, the impedance of the clothesline would be the ratio of the force applied to the line to the resulting transverse velocity. For an electromagnetic wave traveling through a non-magnetic transparent medium, the impedance would be the ratio of the AC electric field to the resulting displacement current. The impedance of a wave medium can be thought of as a measure of the ease or difficulty of getting a wave launched onto the medium.

The input impedance of a wave medium is determined partly by the propagation characteristics of the medium itself and partly by the impedance of the terminating load at the other end. If the medium is infinitely long, or if it has such absorption loss that any reflections returning from the far end are imperceptible at the input, then the impedance of the medium is determined by its own characteristics alone, regardless of the nature of the far-end termination. In this respect we can speak of the characteristic impedance of the medium.

Whenever a traveling wave encounters an abrupt change in the characteristic impedance of the medium it is traveling on or in, the wave speed changes and reflection occurs. The amount of reflection depends on the disparity of the impedances on the two sides of the discontinuity. Thus light is partly reflected in crossing from air to glass. Sound is partly reflected in passing from the low-impedance air column in the vocal tract to the higher-impedance air in front of your face. A wave launched onto a stretched cord will be partly reflected at any place where the cord is attached to a lighter string or to a heavier rope. Impedance transformers are used to smooth over such discontinuities and permit the transmission of wave energy across such boundaries with minimum reflection loss. Elimination of reflection loss is important when only a limited amount of energy is available for transmission and it is desired to deliver as much of it as possible to the energy-consuming load. The designer of a long-distance communication system, having a number of booster amplifiers along the way, must be concerned about possible reflection losses at the interfaces between the line sections and the amplifiers. Not only will such losses reduce the intensity of the signal received by the listener, but also the reflected echoes will confuse the speaker. Similarly, the designer of a multiplet camera lens having several air/glass and glass/glass surfaces must reckon with the reflection losses at all these interfaces. Such losses represent an outright reduction of the light intensity reaching the film and also, by multiple reflections between the lens components, cause a general hazing and a loss of brilliance in the picture.

In cases such as these the impedance mismatch can be smoothed over by inserting a transformer section having a gradual impedance taper between the two impedances being matched. A cheerleader's megaphone (figure 12.9) is an example of a taper transformer. Not only does the megaphone give direction to the sound, but through impedance matching it 'pulls' more sound out of the vocal tract to begin with. In communication technology when it is desired to transmit electromagnetic energy across the junction between two waveguides having different cross sectional dimensions, a tapered waveguide section is inserted at the junction.

Tapered-section transformers are 'broad-band' in that they are effective over a large range of frequencies. However, to be optimally effective, the length of the tapered section must be longer than the wavelength of the longest wave signal to be transmitted. This condition dictates that a cheerleader's megaphone be several feet in length, while microwave tapered-section transformers need be only a few inches long. At 60 Hz power-line frequencies, tapered-section transformers would, of course, have

Figure 12.9 A megaphone is a tapered transformer.

to be hundreds of miles long; hence transformers of the wire-coil type are used exclusively.

Less bulky impedance-matching devices are the quarterwave transformers used at microwave and optical frequencies. A quarterwave transformer is a section of the transmission medium a quarter of a wavelength long, having a characteristic impedance equal to the geometric mean of the impedances of the two media to be matched. The non-reflecting coatings on the lenses of optical instruments are transformers of this type, having a thickness of a quarter of a wavelength and an index of refraction (the optical physicist's term for a quantity related to the electrical physicist's impedance) equal to the geometrical mean of the indices of refraction of air and glass. A quarterwave transformer works because the partial reflections returning from the two ends of the transformer section cancel each other by interference in the backward direction. It is interesting to compare the action of a quarterwave transformer with that of a Fabry–Perot interferometer (see Chapter 9). In the latter device the partially reflecting surfaces are *half* a wavelength apart, resulting in enhanced reflection and reduced transmission.

Compared with tapered-section transformers, quarterwave transformers are 'narrow-band,' in that they give optimum transmission only at frequencies at and close to those for which they are designed. However, in this sense the entire visible spectrum is narrow-band, since it occupies somewhat less than a single octave of the electromagnetic frequency panorama. (Borrowing from music the definition of an octave as an interval between two frequencies one of which is double the other, the range of electromagnetic radiation which we detect visually is about 7.7×10^{14} Hz (violet) to 3.9×10^{14} Hz (red), i.e. a ratio of about 2:1, but in the whole electromagnetic spectrum from radio waves to gamma rays there are about 75 octaves.) A non-reflecting lens coating designed for mid-spectrum green will still be fairly effective in reducing reflection at the blue and red extremities as well.

Well, how did it go? Were you surprised to discover that there is more to the transformer story than the fist-size devices in the power pack of your radio receiver or those wire-sprouting containers sitting on the cross-trees of the power-line poles on your street? Transformers are very much a part of your life. In fact, they're part of *you*! In your middle ear there are linkages of tiny bones which provide an impedance match between the air outside your ear drum and the liquid behind the oval window of your inner ear. Nature must have found this a good impedance-matching device, for it appears in many creatures.

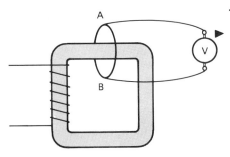

Figure 12.10 If the primary voltage is numerically N_P V, what voltage will the voltmeter indicate when connected as shown to the one-turn secondary?

► **Exercises**

12.1 Describe the kind of wire you would use and the primary–secondary turn ratio for a transformer to operate a small spot-welding machine from a 120 V AC source to give a large current at 3 V across the welding tips.

12.2 Under what conditions is it feasible to operate two transformers in parallel; that is with primaries connected to the same source, secondaries connected to the same load?

12.3 The secondary of the transformer shown in figure 12.10 is a single turn of thick copper wire. The alternating flux in the core produces 1 V per turn. Will an ideal voltmeter connected to diametrically opposite points A and B indicate $\frac{1}{2}$ V, the voltage of half a turn? Explain.

12.4 Figure 12.11 shows an autotransformer, a common type of transformer used for (a) voltage step up or (b) step down. A single continuous winding of wire about an iron core provides both primary and secondary. By means of taps or a slider the turn ratio N_S/N_P and hence the voltage ratio V_S/V_P may be varied.

(a) Show that for comparable ratings the I^2R losses of an auto-transformer are less than those of the conventional type. HINT: In the latter all the load current circulates through the secondary winding.

(b) Show that a disadvantage of the autotransformer as a *power* transformer is that the entire primary voltage may be placed across the load if the secondary winding is accidentally opened.

12.5 One end of a power transmission line of resistance 6 Ω (total) is connected to a 220 V source and the other end to a load resistance of 16 Ω. Find the power consumed by (a) the line and (b) the load.

12.6 If the source voltage of Exercise 12.5 is stepped up to 2200 V by a transformer and then down to 160 V at the load, find the power consumed by (a) the line and (b) the load. Assume 100% efficiency for both transformers.

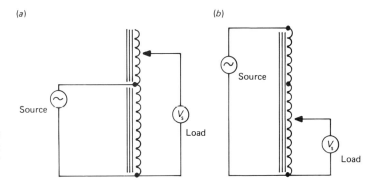

Figure 12.11 In an autotransformer a single continuous winding serves as both primary and secondary to (*a*) step up or (*b*) step down the voltage of a source.

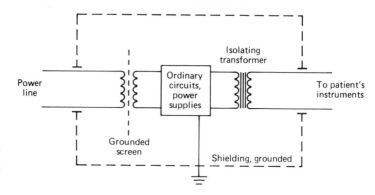

Figure 12.12 An isolation transformer protects a patient from dangerous currents.

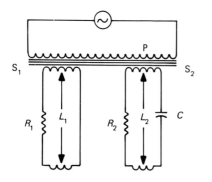

Figure 12.13 A phase-splitting transformer provides currents in each of two output circuits at different phase angles.

12.7   Another application of transformers is the isolation transformer. Figure 12.12 shows a ground-free system for a hospital where all connections to a patient's monitoring instruments are made through an isolation transformer. Show that even if a fault develops there can be no current through the patient to earth. (Since ventricular fibrillation and death can result from a body current as small as 0.1 mA the importance of such isolation should be apparent.)

12.8   If you look at a camera lens which has been given a non-reflecting coating, it may appear to have a faint purplish tint. Explain the tint.

12.9   For a power transformer the voltage ratio V_1/V_2 is called the transformation ratio a. Show that a transformer is capable of transforming circuit impedance values according to the square of the transformation ratio.

12.10   A phase-splitting transformer (figure 12.13) may be used when it is necessary to obtain in two circuits currents which are equal in magnitude but whose phases are in quadrature (at a 90° phase angle). (a) Explain how this is achieved by reference to the diagram. (b) Show that the impedance of the source does not enter into the calculation.

13 Dispersion

Physics tries to discover the pattern of events which controls the phenomena we observe. But we can never know what this pattern means or how it originates; and even if some superior intelligence were to tell us, we should find the explanation unintelligible.

Sir James Hopwood Jeans

13.1 Introduction

Nature presents us with many examples in which the speed of a wave as it travels through a transparent medium depends on the wavelength. Perhaps the best known example is the dispersion of a beam of white light as it travels through a glass prism. Because the index of refraction of the glass is different for different wavelengths, the several monochromatic components of the white light are deviated by different angles, follow different paths through the prism, and are thus separated into a spectrum of colors. Another example is the dispersion of the surface waves of water: one can observe that long-wavelength ocean rollers travel faster than waves of shorter wavelength. In acoustics, the velocity of sound of audible frequencies is markedly independent of wavelength, but at higher frequencies in the ultrasonic range dispersion does occur. Dispersion on electrical transmission lines is a phenomenon of concern to communication engineers.

While the optical physicist refers to dispersion as change of index of refraction with wavelength, $dn/d\lambda$, we shall treat dispersion in the more basic context of change of wave velocity with wavelength, $dv/d\lambda$.

13.2 Observations on Dispersion

Early measurements of optical dispersion for a number of transparent substances showed that the dispersion increases towards the blue end of the visible spectrum. Also, in general, materials with higher indices of refraction tend to exhibit higher dispersions throughout the visible spectrum. These qualitative regularities led to attempts to describe dispersion by mathematical formulae. One of the more successful of these formulae, given by Augustin Cauchy in 1837, was

$$n = A + \frac{B}{\lambda^2} + \frac{C}{\lambda^4},$$
(13.1)

where n is the index of refraction and A, B, and C are constants which can be evaluated for any material by determining the index at three different wavelengths. Differentiation of equation (13.1) to obtain the dispersion $dn/d\lambda$ yields

$$\frac{dn}{d\lambda} = -\frac{2B}{\lambda^3} - \frac{4C}{\lambda^5}.$$
(13.2)

Dispersion which can be described by Cauchy's equation is called *normal dispersion*. If the refractive index of a substance exhibiting normal dispersion at visible wavelengths is followed into the infrared or ultraviolet regions of the spectrum (figure 13.1), extreme departures from Cauchy's formula are found. These departures occur in wavelength ranges where the substance exhibits high absorption and becomes non-transparent. In these regions the dispersion was described by early investigators as *anomalous*. In the light of present-day understanding, anomalous dispersion is just as much a predictable and explainable feature as normal dispersion. However, the term anomalous remains.

While Cauchy's equation was largely empirical, W Sellmeier in 1871 developed an expression, based on the reasoning employed today, which gave for the index of refraction

$$n^2 = 1 + \frac{A_0 \lambda^2}{\lambda^2 - \lambda_0^2} + \frac{A_1 \lambda^2}{\lambda^2 - \lambda_1^2} + \dots, \tag{13.3}$$

where $A_0, A_1, \dots$ are constants and $\lambda_0, \lambda_1, \dots$ are the central wavelengths of the absorption bands.

Sellmeier's equation gives a good representation of the index of refraction everywhere except within the absorption bands. The equation predicts that as λ approaches λ_0 or λ_1 the refractive index should approach infinity. We know that it doesn't. Careful measurement of n within an absorption band can be made, despite the absorption, with thin prisms having very small apex angles. The refractive index within an absorption band goes through a very rapid change with positive slope, as shown by the broken portions of the curve in figure 13.1. Sellmeier's equation was later modified by H L F von Helmholtz, again on the basis of a valid physical model, to give an index of refraction curve which follows exactly those found experimentally at all wavelengths.

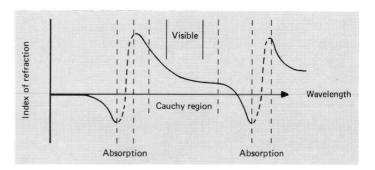

Figure 13.1 The index of refraction depends on wavelength.

13.3 The Physics of Dispersion

We wish now to investigate the physical basis for the dependence of wave velocity upon wavelength. In particular we seek an explanation of the extreme variations occurring in the absorption regions of the spectrum.

When an atom is placed in an electric field its valence-electron shell is pulled aside by a small displacement so that its center no longer coincides with the center of the atom. The atom thus becomes a dipole. Such is the case with the atoms in a plane layer of atoms pictured edge-on in figure 13.2, with a train of plane electromagnetic waves passing through it. Since the electric field to which the atoms are subjected is alternating, the dipole moment of each atom in the plane is alternating, and the atoms behave like oscillators driven in forced oscillation by the alternating field of the incoming waves. Because an alternating dipole radiates electromagnetic energy, each atom in the plane becomes a source of secondary wavelets.

These secondary wavelets from the atoms in the plane cancel each other by destructive interference in all directions except the original forward direction. In the forward direction the wavelets combine to produce secondary waves which travel off to the right along with what remains of the original primary waves. While the secondary waves have the same frequency as the primary waves, their amplitude will depend upon the number of atoms per unit

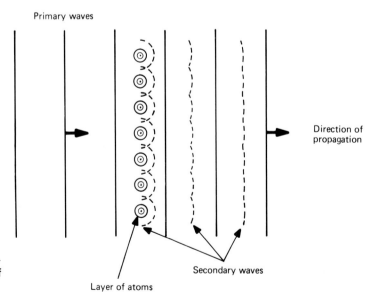

Primary waves

Direction of propagation

Secondary waves

Layer of atoms

Figure 13.2 Secondary waves are produced by the forced oscillation of electron-shell oscillators.

area in the plane and on the 'spring constant' of the electron shells. Their phase will depend upon how the driving frequency of the incoming wave relates to the natural frequency f_0 of the electron-shell oscillators. The resultant waves emerging from the plane of atoms will be the sum of the primary and secondary waves, with due regard for their relative amplitudes and phases. These resultants then become the primary waves for the next plane of atoms. The absorption characteristics and the index of refraction of the medium depend upon the amplitude and phase of the secondary waves. Let us see how this dependence comes about.

When an electromagnetic wave passes over a 'parasitic' oscillator (one driven by an external source), such as an atom with a displaceable electron shell, the phase of the secondary waves produced lags 90° behind the phase of the displacement of the electron shell. As we saw in §6.5 on forced oscillators, this displacement can lag by anywhere from 0 to 180°, depending upon the frequency of the driver. The secondary waves produced by the plane of atoms, therefore, can have a phase lag anywhere between 90° and 270° behind the phase of the primary waves. The amplitude of the secondary waves depends on how close the frequency of the primary waves is to the natural frequency f_0 of the electron-shell oscillators.

At frequencies less than f_0 the phase of the secondary waves lags between 90° and 180° behind that of the primary waves. This lag, progressively compounded at every succeeding plane of atoms, is interpreted as a retardation of the primary-wave speed and results in an effective resultant-wave velocity somewhat less than c, the wave velocity in free space. At frequencies higher than f_0, the phase of the secondary waves lags between 180° and 270° behind that of the primary waves. This lag can be thought of as a phase advancement of the next following primary wave and results in an effective resultant-wave velocity greater than c. At and near the resonant frequency of the electron shells the amplitude of the secondary waves becomes very large at the expense of the amplitude of the primary waves. Moreover, the secondary waves are at or near phase opposition to the primary waves, which therefore become progressively damped as they proceed through layer after layer of atomic oscillators. In this frequency region the absorption of electromagnetic energy is large; the material is opaque.

The natural frequencies of the electron-shell oscillators of most transparent substances lie in the near ultraviolet region of the spectrum. With this generalization in mind, we can account for the index of refraction and the absorption coefficient of a 'typical' transparent substance in the four wavelength ranges described below and depicted in figure 13.3.

Range I, far infrared The primary-wave frequency is very much less than f_0. The secondary waves lag slightly more than 90° behind the primary waves. The primary waves, as they travel through the medium, are progressively retarded in phase by compounding with the secondary waves to which they give rise. The resultant apparent wave speed is less than c. The electron oscillators, being excited far from their natural resonance frequency, respond with small amplitude at low frequency and absorb almost no electromagnetic energy, so the material is transparent.

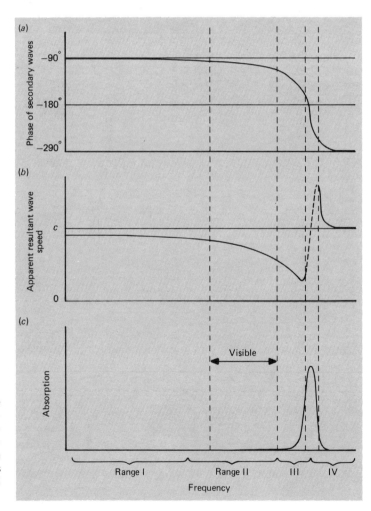

Figure 13.3 (*a*) As the frequency of the primary waves goes through an absorption region, the phase lag of the secondary waves shifts from 90° to 270°. (*b*) The resultant-wave speed varies with frequency. (*c*) The absorption maximizes at the frequency of electron-shell resonance.

Range II, near infrared and visible With increasing frequency the secondary waves become larger in amplitude and their phase lag moves from 90° towards 180°. Their retarding effect on the primary waves becomes appreciable and the resultant apparent wave speed decreases. The amplitude of the electron-shell oscillators is still too small to account for much absorption. The material continues to be essentially transparent.

Range III, near ultraviolet With the frequency approaching the resonance frequency of the atomic oscillators, the amplitude of the oscillation becomes very large, as does the amplitude of the secondary waves. The phase lag of the secondary waves approaches 180°, in opposition to the phase of the primary waves, which progressively decrease in amplitude as they penetrate farther into the medium. At resonance the absorption of electromagnetic energy by the oscillators is maximum, and the material is opaque. The absorbed energy is communicated to the crystal lattice and appears as a heating of the medium.

Range IV, frequency of the primary wave increasing on the high-frequency side of the electron-shell resonance As the exciting frequency increases further, the amplitude of the oscillation falls off, as does the amplitude of the secondary waves. The phase lag of the latter increases from 180° towards 270°, with the phase of the resultant waves being advanced. The apparent resultant-wave speed is greater than c. As the amplitude of the secondary waves decreases towards zero with increasing frequency in this range, the magnitude of the advancement decreases, and the resultant-wave speed approaches c at very high frequencies. It must not be inferred that the speed of the primary waves becomes greater than the speed of light. The primary waves progress through the medium with precisely the speed of light in free space, but the resultant of the primary waves, the secondary waves they produce as they pass over layer after layer of atoms, and the snowballing family of additional secondaries produced by these secondaries, does propagate with a velocity greater than the velocity of light in free space. This velocity is called *phase velocity* and is the velocity with which a crest of the resultant wave proceeds. It is the effective velocity which is measured in a deviation experiment with a prism.

The speed with which information can be transmitted through a medium can never exceed the speed of light in free space. Information can be carried by waves only by impressing upon them some change of character, such as turning them on and off or altering their amplitude or frequency. Such a change does not travel with the phase velocity but at a slower rate called the signal velocity or group velocity.

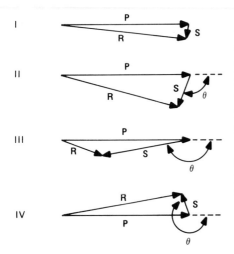

Figure 13.4 As the phase and magnitude of the secondary waves change with wavelength, so do the phase velocity and absorption of the resultant waves. **P**, primary; **S**, secondary, **R**, resultant.

The phase relationships between the primary and secondary waves in these frequency ranges can be presented in the form of vector diagrams (figure 13.4). In each of these diagrams the primary wave is represented by a vector of constant magnitude directed along the axis of positive real numbers. The secondary wave is then added, as a vector of appropriate magnitude and phase. From the resultant's magnitude and phase the absorption and phase velocity can be deduced.

When these ideas are put into mathematical form one obtains an expression for the index of refraction n in terms of the properties of the atoms of the material and the frequency of the radiation:

$$n^2 = 1 + \frac{Ne^2}{8\pi^2 \varepsilon_0 m (f_0^2 - f^2)}, \qquad (13.4)$$

where N is the number of atoms per unit volume, e the electron charge, ε_0 the permittivity of free space, m the electron mass, f_0 the frequency of the electron-shell resonance, and f the frequency of the radiation. The right-hand side of this equation may be recognized as being essentially the same as the first two terms of Sellmeier's equation (13.3).

While we have so far developed the physics of dispersion in terms of a single mechanism (electron-shell oscillation) there are other mechanisms which contribute to the index of refraction. In a transparent ionic solid, such as rock salt, there are charged ions which execute forced vibrations under the influence of the incoming primary waves. For most ionic substances these ionic vibrations have resonances in the infrared spectral region, giving rise to absorption bands and ranges of anomalous dispersion in the infrared. On the long-wavelength side of the ionic absorption region of glass, the index of refraction is around 3, whereas on the short-wavelength side (including the visible region) the index is only 1.5 to 1.7.

Still other substances, particularly liquids and gases, have polar molecules which can rotate so as to keep their dipole axes aligned with the alternating electric field of the incident radiation. At low frequencies this alignment can easily keep up with the alternations of the radiation field. However, as the frequency f is increased, the frictional opposition to molecular rotation begins to absorb energy from the radiation and the material becomes increasingly opaque. This absorption maximizes at a frequency where anomalous dispersion occurs, usually in the far-infrared or short-microwave regions. On the high-frequency side of this absorption band the molecules are too sluggish to rotate significantly and the material again becomes transparent. Water is a good example of a polar substance exhibiting this behavior. Its index of refraction at visible frequencies is 1.33, while from radio frequencies to zero frequency, (where molecular rotation is fully effective) its index of refraction is around 9.

When these additional mechanisms are taken into account mathematically, additional terms must be included in equation (13.4) to represent the contributions to the index of refraction made by ion vibration and molecular rotation. These terms will involve f_1 and f_2, the central frequencies of the absorption bands due to these processes.

According to classical electromagnetic theory the index of refraction of a transparent non-magnetic substance is equal to the square root of the dielectric constant. The latter is, of course, a measure of the extent to which a sample of the material can be polarized when subjected to an electric field. As we have seen, the various physical mechanisms contributing to polarization, namely electron-shell displacement, ion displacement, and polar molecule alignment, are frequency dependent. They are all effective at very low frequencies, leading to the high dielectric constants measured in static and low-frequency experiments. With increasing frequency the molecular alignment contribution disappears first. At still higher frequency the ionic vibration contribution disappears. Finally, at some ultraviolet frequency even the electron-shell vibration can no longer keep up with the oscillation of the impressed radiation field, and the material is left with a dielectric constant essentially equal to 1.0 in the far-ultraviolet and soft x-ray regions. In the harder x-ray region (lower frequency/higher frequency) the dispersion curve goes through some additional ups and downs owing to inner-shell electron resonances which we shall not dwell upon here. Figure 13.5 shows how the dielectric constant of a hypothetical

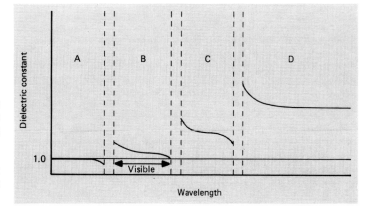

Figure 13.5 The dielectric constant is made up of contributions from various physical mechanisms in different frequency ranges. Region A, nothing; B, electron-shell displacement; C, electron-shell displacement + ion vibration; D, electron-shell displacement + ion vibration + molecule alignment.

'typical' substance depends upon frequency in the way we have just discussed.

13.4 Group Velocity

If all the simple waves which make up a group travel with the same velocity, the group will move with this velocity and will maintain its shape unchanged, but if the velocities vary with wavelength the group will change its form as it progresses. Water waves are a familiar example. When you throw a pebble into the water and watch the ripples spread out in all directions, you may observe that the group of waves travels with a velocity less than that of any particular wave crest in the group. As you watch, you see new wave crests appearing out of nowhere at the rear of the group, traveling forward within the group, and disappearing out in front of the group. As we have already discussed, the velocity of any particular wave crest is the phase velocity while the velocity of the packet of crests as a whole is the *group velocity*.

The reason for the difference between phase and group velocity lies in the dispersion of the water surface as a medium for wave propagation. When the stone is thrown into the water, the water at the place of impact oscillates up and down a few times, emitting a train of waves having not a single wavelength but rather a distribution of different wavelengths within a narrow band. The different wavelength components of this composite packet travel with different velocities leading to the superposition behavior you observe.

The analysis of group velocity can be undertaken easily by supposing that two discrete wavelengths λ and $\lambda + \Delta\lambda$ have the same amplitude and travel with individual velocities v and $v + \Delta v$. These two wave trains are pictured in figure 13.6. Their superposition at any instant produces beats.

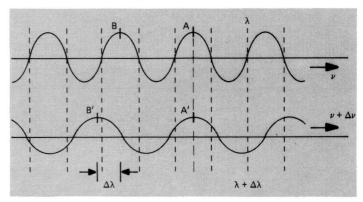

Figure 13.6 Crest B' of the lower wave train catches up with crest B of the upper wave train in time $\Delta\lambda/\Delta v$.

Let us start counting time at the instant that the crests A and A′ of the two wave trains are exactly superposed. This superposition corresponds to the peak in the envelope of a beat. Now let both wave trains run until crest B′ catches up with crest B. The time required for this overtaking is $\Delta\lambda/\Delta v$. In this time the two wave trains will have advanced a distance $v\Delta\lambda/\Delta v$. However, the crest of the beat envelope will have fallen back one wavelength, from AA′ to BB′. The beat envelope thus travels with velocity

$$v_{\text{env}} = \frac{v\Delta\lambda/\Delta v - \lambda}{\Delta\lambda/\Delta v} = v - \lambda\frac{\Delta v}{\Delta\lambda}, \qquad (13.5)$$

which is the group velocity of the beat packet. Note that if $\Delta v/\Delta\lambda$ is positive (as it is for water surface waves) the group velocity is less than the phase velocity. Should the reverse be true, the group velocity would be greater.

In any experiment for determining the velocity of light by timing the progress of spurts of light produced by a shutter of some kind, the velocity measured is the group velocity. A light beam is truly monochromatic only if it consists of an infinitely long train of waves. Shuttering the beam on and off produces a Fourier continuum of different wavelengths grouped about a central wavelength, traveling with different velocities and exhibiting the group velocity phenomenon. Fizeau's and Michelson's velocity measurements were thus measurements of group velocity.

There is no dispersion for light traveling in a vacuum, where the group velocity is equal to the phase velocity. The most precise evidence for this statement comes from observation of the light from eclipsing binary stars. For Algol, 120 light years† distant, eclipses occur at intervals of 68 h 49 min. If v_{red} differed from v_{blue} by as little as 1 part in 10^6 the starlight would show a measurable time delay in the eclipse for different colors. It doesn't.

In air the group velocity of light is less than the phase velocity by only 0.007% (or $2.2\,\text{km s}^{-1}$). In water the difference is 1.5% and in crown glass about 2.4%.

13.5 Whistlers

During World War I while Heinrich Barkhausen was eavesdropping on Allied field-telephone conversations at a distance, he occasionally heard curious whistling sounds which swamped

† 1 light year = 6×10^{12} miles.

the military messages. Unable to eliminate the whistles from his apparatus, he concluded that they must come from the atmosphere.

In the 1920s and 1930s investigators at Bell Laboratories and at Marconi Wireless Telegraph noted that a whistler often appeared about a second or so after a loud atmospheric click. They developed the explanation that when lightning occurs, the very low-frequency components of the discharge are propagated along lines of the Earth's magnetic field to the antipodes. The click is a rectangular wave-form composed of many different wavelengths, for the same click can be detected all over the broadcasting band. Now suppose that as a click moves through the ionosphere, its component frequencies will spread out, the highest frequency traveling fastest and the lower ones strung out behind. If the click travels far enough before the magnetic lines guide it to a receiver, the frequencies will be well separated. The observer will receive a drawn-out signal—a whistling tone of steadily falling pitch. Whistlers are an interesting example of nonlinear propagation leading to dispersion.

13.6 Solitons

As physicists, we have experienced a 'linear' upbringing, via the harmonic oscillator and Schrödinger's equation, but nature also presents large-amplitude structures which do not spread in time. J Scott-Russell is credited with observing the first soliton when in 1834 he observed in a canal in Scotland the remarkable stability of a solitary wave in shallow water and the break up of an initial pulse into two solitons. In 1965 N J Zabusky and M D Kruskal, using an anharmonic crystal model, found wave solutions and showed that solitary-wave solutions were extremely stable on collision.

Solitary-wave (soliton or kink) solutions of the equations of motion of constituents of a system undergoing a phase transition have led to experimentally verified results. By devising methods for solving nonlinear evolution equations, mathematicians have provided a soliton paradigm useful in numerous areas of nonlinear physics, such as lattice dynamics, phase transitions, nonlinear dispersive systems, grain boundaries, and nonlinear optics, plasmas, and hydrodynamics.

► **Exercises**

13.1   Can you think of any basis for the occasionally-heard statement that every seventh ocean wave is larger than the other six?

13.2   Why should water have a DC dielectric constant (81) so much larger than that of benzene (4)?

13.3   Many substances have more than one absorption band in the infrared region of the spectrum. Can you explain why?

13.4   If there is a vacuum inside a wave guide, show that the phase velocity v of electromagnetic waves in the wave guide is greater than the velocity of light c in free space.

13.5   Show that equation (13.4) can be put into the form of the first two terms of Sellmeier's equation.

13.6   Show that for water at wavelength 5893 Å the group velocity of light waves is 1.5% less than the phase velocity. Take $dn/d\lambda = 3.38 \times 10^{-6} \text{ Å}^{-1}$.

13.7   Find the group velocity u by considering the pulse consisting of a superposition of two harmonic waves, $y = \cos(\omega t - kx)$ and $y = \cos(\omega' t - k'x)$, where ω is angular frequency, $\omega = 2\pi f$.

13.8   Find expressions for the group velocity $u = d\omega/dk$ (a) in terms of the phase velocity v and dispersion $dv/d\lambda$ and (b) in terms of v and $dn/d\lambda$.

13.9   Find the group velocity for situations when the following dispersion laws apply (v is the phase velocity):

(a) $v = a = $ constant,   sound waves in air
(b) $v = a\sqrt{\lambda}$,   gravity waves on the surface of water
(c) $v = a/\sqrt{\lambda}$,   capillary waves
(d) $v = a/\lambda$,   transverse vibration of a rod
(e) $v = \sqrt{(c^2 + b^2\lambda^2)}$,   electromagnetic waves in the ionosphere.

13.10   The dispersive power of glass for visible light may be defined as the ratio $n_D/(n_C - n_F)$, where C, D, and F refer to Fraunhofer lines in the red ($\lambda_C = 6563$ Å), yellow ($\lambda_D = 5893$ Å) and blue ($\lambda_F = 4861$ Å). Find the approximate group velocity in glass which has dispersive power 30 and $n_D = 1.5$.

13.11   Show that if the index of refraction were represented *exactly* by the first two terms of equation (13.1), it would be possible to construct a lens free from chromatic aberration by combining a positive lens and a negative lens made of different kinds of glass. HINT: recall that for a lens with spherical surfaces the focal length f is related to the curvatures r by

$$\frac{1}{f} = (n-1)\left(\frac{1}{r_1} + \frac{1}{r_2}\right).$$

14 The Ubiquitous *kT*

A physical theory is not an explanation; it is a system of mathematical propositions whose aim is to represent as simply, as completely, and as exactly as possible a whole group of experimental laws.

Pierre Duhem

14.1 Introduction

You can't get very far in your study of physics without running into a formula or an equation containing the combination of symbols kT, where k is Boltzmann's constant and T is absolute temperature. Indeed, the deeper you penetrate, the more frequently you encounter this combination, until it may seem as if a large part of physics rests on this quantity in some way. It appears again and again as you study the gas laws, kinetic theory, thermal and electrical conduction, diffusion, chemical processes, thermionic emission, noise, and a host of other topics. Although kT had its genesis in classical physics, it survived the transition to quantum physics and it is with us today as vigorous as ever.

It will be the purpose of this chapter to elucidate the significance of the product kT and to give a number of examples of its applicability in situations where the statistical behavior of atoms, molecules, or electrons determines the physical process being studied.

14.2 Kinetic Theory

Your first encounter with the ubiquitous product kT most likely occurred when you examined the kinetic-theory model of an ideal gas. You followed a simple derivation for the pressure p of a volume V of gas containing N molecules, resulting in the expression

$$pV = \tfrac{1}{3}Nmv^2, \qquad (14.1)$$

where m is the mass of the individual molecule and v is its RMS thermal velocity.

You further followed the experimental establishment of the three ideal gas laws inter-relating the pressure, volume, and temperature of a quantity of confined gas:

$$\frac{p_1}{p_2} = \frac{V_2}{V_1} \quad \begin{array}{l}\text{at constant temperature} \\ \text{(Boyle's law)}\end{array}$$

$$\frac{p_1}{p_2} = \frac{T_1}{T_2} \quad \begin{array}{l}\text{at constant volume} \\ \text{(Charles's law)}\end{array} \qquad (14.2)$$

$$\frac{V_1}{V_2} = \frac{T_1}{T_2} \quad \begin{array}{l}\text{at constant pressure} \\ \text{(Gay-Lussac's law)}.\end{array}$$

You then pursued Boyle's law, which can be rewritten as $pV = c$, where c is a constant, through an experimental determination of c, which has different values at different temperatures. This led to the result

$$pV = nRT, \qquad (14.3)$$

where n is the number of gram molecular weights (moles) of gas present and R is the universal gas constant ($8.31\,\mathrm{J\,mole^{-1}\,K^{-1}}$).

From here we take off into what may be for you new territory. Combine equations (14.1) and (14.3) to obtain $\frac{1}{3}Nmv^2 = nRT$. If we consider a sample of gas containing exactly one mole, then n is equal to 1.00 and N becomes the Avogadro number N_A (the number of molecules in one mole). Then $\frac{1}{3}mv^2 = (R/N_A)T$. The fraction R/N_A may now be replaced by the quantity k, which has become known as the Boltzmann constant and has the value $1.38 \times 10^{-23}\,\mathrm{J\,K^{-1}}$. The last equation then becomes

$$\boxed{\tfrac{1}{3}mv^2 = kT,} \qquad (14.4)$$

and here we have the first appearance of the product kT in this book.

What does kT mean and why does it appear so frequently in physical relationships? If we multiply both sides of equation (14.4) by 3/2 we obtain

$$\tfrac{1}{2}mv^2 = \tfrac{3}{2}kT. \qquad (14.5)$$

Immediately we recognize the left-hand side of this equation as the kinetic energy of thermal motion of a molecule as it moves about, colliding with other molecules and with the walls of the enclosure. Hence $\frac{3}{2}kT$ must have the physical significance of being equal to the thermal kinetic energy of a molecule.

▶ **Example 14.1** A one liter spherical flask contains hydrogen (molecular diameter 2.2×10^{-10} cm) at 300 K. What must the pressure be so that the mean free path (MFP) of a molecule is greater than the diameter of the flask?

The mean free path (MFP) l is the average distance a molecule will travel before colliding with another molecule. For a molecule of diameter D, imagine a cylinder of radius D and length l centered around its path. If the center of any other molecule comes within this cylinder, there will be a collision. Thinking this way we can get an estimate of the relation between the MFP l and the diameter D of a molecule. The volume of the cylinder is $\pi D^2 l$. The probability that another molecule will be in it is proportional to the number N of molecules in the enclosure and inversely proportional to the available volume V. Thus,

$$l = \frac{V}{\pi D^2 N}$$

From equation (14.1)

$$\frac{V}{N} = \frac{1}{3}\frac{mv^2}{p},$$

but $\frac{1}{3}mv^2 = kT$ from equation (14.4),

$$\therefore \frac{V}{N} = \frac{kT}{p},$$

$$\therefore l = \frac{kT}{p\pi D^2}.$$

For the flask $V = 10^{-3}\,\mathrm{m}^3 = 4\pi r^3/3$. Therefore $r = (10^{-3} \times 3/4\pi)^{1/3}$ $= 0.0625\,\mathrm{m}$, and the diameter is $0.125\,\mathrm{m}$. For l to be greater than this diameter,

$$\frac{kT}{p\pi D^2} > 0.125,$$

$$\frac{kT}{0.125\pi D^2} > p.$$

Thus,

$$p < \frac{1.38 \times 10^{-23} \times 300}{0.125\pi \times (2.2 \times 10^{-10})^2},$$

$$p < 0.219\,\mathrm{N\,m}^{-2} = 2.1 \times 10^{-6}\,\mathrm{atm}.$$

At this point you may rebel. 'Why the three halves?' you complain. 'Why don't you give a numerical value for k that will make kT itself, rather than $\frac{3}{2}kT$, equal to the thermal energy of a molecule?' Let's see if we can rationalize this.

The molecule we're talking about is continually changing its speed and its direction of motion, depending on the speed and direction it had before its last collision and on the speed and direction of the last thing it collided with. Consequently the RMS velocity v represents a long-time average of many instantaneous random speeds and directions. In any random direction at any instant

$$\tfrac{1}{2}mv^2 = \tfrac{1}{2}mv_x^2 + \tfrac{1}{2}mv_y^2 + \tfrac{1}{2}mv_z^2 = \tfrac{3}{2}kT,$$

where v_x, v_y, and v_z are the RMS components of the velocity along the x, y, and z directions. On the long-time average v_x, v_y, and v_z are equal and hence

$$\tfrac{1}{2}mv_x^2 = \tfrac{1}{2}mv_y^2 = \tfrac{1}{2}mv_z^2 = \tfrac{1}{2}kT.$$

Therefore we see that $\frac{1}{2}kT$ represents the average thermal kinetic energy of a molecule associated with each direction of three-dimensional space, or, as we say, with each degree of freedom of translational motion.

With this interpretation of what is meant by kT, it is easy to see why kT is so important a quantity in the scheme of things. Since so many physical processes involve the thermal energies of individual molecules, atoms, or charged particles, it is no wonder that one finds kT sprinkled so liberally through the formulae and expressions describing things that happen in physics. Without attempting to be comprehensive, let's turn to a few illustrative examples.

14.3 Distribution of Molecular Velocities

In the elementary derivation of equation (14.1) it was assumed that all the particles had the same thermal velocity and that, on average at any instant, one third of the particles were moving in the x direction, one third in the y direction, and one third in the z direction. Neither of these assumptions is correct, but a more rigorous analysis, taking into account both the spread of particle velocities and the randomness of their directions of motion, leads to the same result because we are dealing with a very large number of molecules and hence statistical arguments can validly be used.

We wish now to see how the magnitudes of thermal velocities of molecules are distributed. First, consider a simple model. If we take two dozen marbles, place them in a tray, and shake the tray, we will see at once that the marbles take on a wide range of velocities in all directions. They collide with each other and with the sides of the tray, changing velocities and directions at every collision. However, there is a stability in this chaos. While some marbles are moving slowly at any particular instant and others are moving much faster, it is possible to discern an average velocity which remains quite stable at any constant level of shaking. If the tray is shaken more violently, the tempo of the chaos increases and quickly assumes a new statistical steady state with a higher average velocity.

Should it not be likewise in a three-dimensional population of gas molecules in a container? The molecular velocities should cover a considerable range from the highest to the lowest, and a molecule that is traveling along at a mile per second at one instant may collide with another molecule and find itself almost stationary the next instant, only to be clipped from the side in the next microsecond and knocked along at a half a mile per second in some new direction. Although the velocity of any individual molecule has its ups and downs, the population as a whole should be characterized by some kind of average speed which remains constant with time.

The task of deducing the distribution of molecular velocities was undertaken by J C Maxwell and by L Boltzmann about a hundred years ago. By different pathways of statistical reasoning they arrived at the same expression:

$$n = 4\pi N \left(\frac{m}{2\pi kT}\right)^{3/2} v^2 e^{-mv^2/2kT} dv, \tag{14.6}$$

where n is the number of molecules in the velocity range dv wide and centered at velocity v, N is the total number of molecules in the enclosure, and m is the mass of one molecule. This equation is known as the Maxwell–Boltzmann distribution. Note that the distribution is independent of the pressure of the gas and the volume in which it is confined. The only variables are the temperature and the molecular mass of the particles.

This equation may not tell you much, just to look at it, so let's consider its plot. Figure 14.1 shows the shape of the velocity distribution for a population of 1000 hydrogen molecules at a temperature of 300 K. The ordinate values are calculated for velocity ranges 10^4 cm s^{-1} wide. The distribution function peaks at a velocity value v_p which, by definition, is the *most probable* velocity, that is, more of the molecules have velocities at or near this value than any other. However, because the curve is not symmetrical, this velocity is not quite the same as the *average* velocity v_a or the RMS velocity referred to in equation (14.1). It can

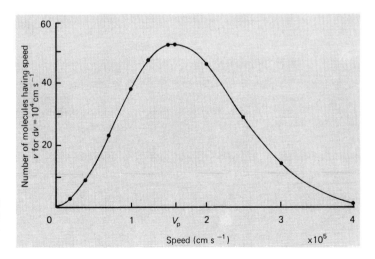

Figure 14.1 The molecules of a gas have thermal speeds shown by the Maxwell–Boltzmann distribution. V_P = 1.58×10^6 mm s^{-1}.

be shown that:

$$\text{average velocity} = 1.128\, v_{\text{p}},$$
$$\text{RMS velocity} = 1.224\, v_{\text{p}}. \tag{14.7}$$

The significance of kT in relationship to the parameters of this distribution curve should be pointed out. Differentiating the right-hand side of equation (14.6) with respect to v and setting the result equal to zero, we can determine that the peak of the curve comes at velocity $v_{\text{p}} = \sqrt{(2kT/m)}$. Thus, for molecules traveling at this most probable velocity, the kinetic energy $\frac{1}{2}mv_{\text{p}}^2$ is equal to kT. This value is to be compared with the kinetic energy of $\frac{3}{2}kT$ for a molecule traveling with RMS velocity, as given by equation (14.5).

If we increase the temperature of the sample of gas, the velocity distribution broadens out, with the peak shifting to higher velocities. If we reduce the temperature or substitute molecules of greater molecular mass, the peak of the curve shifts toward lower velocities. The area under the curve must remain constant, however, since we are talking about a fixed number N of molecules.

By inspection of this curve one can see that approximately 5% of the molecules in the sample will have velocities equal to or greater than twice the most probable velocity, while fewer than 0.1% have velocities in excess of three times the most probable velocity.

▶ **Example 14.2** Calculate the pressure p as a function of height above sea level h for the Earth's atmosphere treated as an ideal gas in a gravitational field.

The pressure p at height h is less than pressure $p + dp$ at height $h - dh$ by the weight of air in a vertical column of unit cross sectional area and of height $dh : dp = -\rho g\, dh$. The density of air is $\rho = pM_0/RT$, where M_0 is 28 g mole^{-1}. Therefore,

$$dp = -\frac{pM_0\, g\, dh}{RT},$$

$$\frac{dp}{p} = -\frac{M_0\, g\, dh}{RT}.$$

On integrating we obtain,

$$\ln p = -\frac{M_0 g h}{RT} + c.$$

When $h = 0$, $p = p_0$ (the pressure at sea level). Therefore $c = \ln p_0$ and we have

$$\ln p = \frac{-M_0 gh}{RT} + \ln p_0,$$

$$p = p_0\, e^{-M_0 gh/RT} = p_0\, e^{-mgh/kT}.$$

This equation is called the barometric equation or the law of atmospheres.

14.4 Specific Heat Capacity of Gases

Specific heat is defined as the quantity of heat (energy) required to raise the temperature of unit mass of a substance by one degree. In many cases we find it more revealing to talk about the molar heat capacity, which is the heat required to raise the temperature of one mole of the substance by one degree. In the latter case we are dealing with the same number of molecules and can make more meaningful comparisons among different substances.

When dealing with the specific heat capacities of gases, it is necessary to distinguish between two different kinds of specific heat capacity. In measuring specific heat capacity at constant volume, c_v, the sample of gas is enclosed in a rigid container which does not expand as the gas is heated and its pressure increases. In measuring specific heat capacity at constant pressure, c_p, the gas, initially at, say, atmospheric pressure, is enclosed in a chamber whose volume can be increased as the gas is heated so that the pressure of the gas remains constant. It should be clear that c_p will be larger than c_v by the additional amount of heat required to do the work of expanding the gas against atmospheric pressure by the volume increment dV. That is, $c_p = c_v + p\,dV$.

Returning to c_v, we recognize that heating a gas is a process whereby we increase the average kinetic energy of the molecules. For one mole the thermal energy $E_k = \frac{3}{2}RT$. The molar heat capacity at constant volume is the derivative of this quantity with respect to temperature:

$$c_v = \tfrac{3}{2} R. \tag{14.8}$$

This expression is found to be substantially correct for gases known to have monatomic molecules. Molar heat capacities for polyatomic gases, however, are found to be considerably larger. This is because these gases have additional degrees of freedom over and above the three simple translational degrees of freedom available to the molecules of a monatomic gas. The molecules of a diatomic gas, for example, can rotate and vibrate as well as move bodily in the *x*, *y*, and *z* directions of space. Consider a molecule of diatomic gas (figure 14.2). The atoms may be considered as mass points joined by a spring representing the elastic binding force between them. If this assemblage is aligned along the *x* axis, for example, it can vibrate along the *x* direction, with the two atoms moving in opposite phase. The energy of this vibration is partly potential and partly kinetic. The assemblage can also rotate about the *y* and *z* axes, but not about the *x* axis since the atoms are regarded as mass points. The diatomic molecule then has three degrees of freedom of translation in space, two degrees of freedom of vibration (potential and kinetic), and two degrees of freedom of rotation, seven degrees of freedom in all. When heat is absorbed by such a gas, the energy is partly channeled into increasing the vibrational and rotational energies of the molecules, leaving only part of the input heat to increase the energy of the translational motions which we interpret as temperature. Hence such a gas will require more input heat than a monatomic gas would require in order to raise its temperature by one degree.

It is a basic postulate of the kinetic theory that the internal energy of a system of particles having several degrees of freedom is, on average, equally divided among those degrees of freedom, each possessing energy $\frac{1}{2}kT$ per particle. This postulate is called the equipartition law. According to this picture, then, the specific heat capacity c_v of a diatomic gas having seven degrees of freedom should be $\frac{7}{2}R$ per mole. It is found experimentally, however, that most diatomic gases around room temperature have molar heat capacities nearer to $\frac{5}{2}R$!

The resolution of this upsetting disagreement did not come about until the introduction of quantum concepts in the early twentieth century. Briefly, this is the story. A molecule cannot vibrate at just any amplitude and energy. The vibration is *quantized*. That is, the molecule is permitted to vibrate at only those amplitudes for which the vibrational energy E is given by $E = (n + \frac{1}{2})hf$ where n is 0, 1, 2, 3, . . . , h is a universal physical

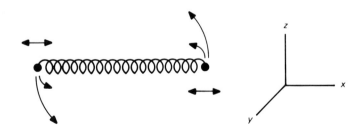

Figure 14.2 A diatomic molecule can vibrate along the *x* direction and rotate about the *y* and *z* directions.

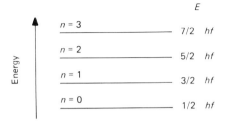

Figure 14.3 The allowed vibrational energy levels of a diatomic molecule are quantized.

constant called Planck's constant, and f is the frequency of the vibration, determined by the masses of the atoms and the elasticity of the forces bonding them together in the molecule. All other amplitudes and energies are forbidden. Figure 14.3 shows the allowed vibrational states of such a molecule.

At very low temperatures all the molecules in the sample are in the lowest possible vibrational state for which $n = 0$. It is interesting that, even at absolute zero temperature, where all molecular motion is supposed to be totally dead, this theory predicts a certain rock-bottom minimum amount of vibration having energy $\frac{1}{2}hf$ per molecule, the 'zero-point vibrational energy.' In order to vibrate with any larger amplitude and energy, the molecule must absorb energy of *at least hf* to transfer it to the vibrational state $n = 1$. For this to occur, the temperature must be high enough so that kT is a substantial fraction of the required energy hf. Otherwise the molecules will continue in vibrational state $n = 0$ and won't accept any vibrational energy with increasing temperature. With the vibrational degrees of freedom thus excluded from participation, the low-temperature specific heat capacity of the gas will remain at $\frac{5}{2}R$ per mole.

With increasing temperature, the ratio of kT to hf eventually becomes large enough to permit the activation of some of the molecules across the energy gap from $E = \frac{1}{2}hf$ to $E = \frac{3}{2}hf$ with the consequent absorption of heat energy into the vibrational motion of the molecules. Not until then will the molar heat capacity begin to rise from $\frac{5}{2}R$ towards $\frac{7}{2}R$. As the temperature is raised still farther, more and more of the molecules will be promoted to the vibrational states $n = 1, 2, 3, \ldots$ with the further absorption of heat energy. When the temperature is reached at which the first half dozen vibrational states become populated with molecules, the distribution of molecules among these states comes to resemble a 'normal' distribution, and only then will the gas exhibit the full $\frac{7}{2}R$ molar heat capacity predicted by the earlier theory.

14.5 Specific Heat Capacity of Solids

The atoms of a solid do not possess kinetic energy of translation, for they do not move freely in the x, y, and z directions, but are anchored to their lattice positions. Neither do they rotate, but they do vibrate in the x, y and z directions. The energy of this vibration is statistically partitioned equally between kinetic and potential energy. The vibrating atoms therefore have six degrees of freedom. Hence, with $\frac{1}{2}kT$ of energy per atom per degree of

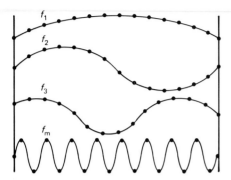

Figure 14.4 A line of atoms in a solid can vibrate in a large number of different modes, each with its own particular energy.

freedom, a specimen of an atomic solid containing one mole of atoms should have a total vibrational energy content of $3N_A kT$ and molar heat capacity $3R$.

The molar heat capacities of many monatomic solids do, indeed, come close to this value at room temperature. This fact was noted experimentally by Dulong and Petit in 1819, long before any theory of specific heat capacity existed. However, at lower temperatures the molar heat capacities of all solids decrease and tend toward zero as the absolute zero of temperature is approached.

This behavior is accounted for by the application of quantum ideas to the vibrations of the atoms of a solid. Suppose we consider an atom in a row of atoms stretching from one side of our specimen to the other in the x direction. This atom can vibrate in an enormous number of different modes, with the line of atoms of which it is a part behaving much like a vibrating string (figure 14.4.) Our chosen atom will, of course, be a unit in other lines of atoms extending through the specimen in the y and z directions as well. Each vibration mode i has a frequency f_i and, according to quantum theory, an energy $E_i = hf_i$. The total vibrational energy of all the atoms of the specimen can be obtained by summing the energies of all the vibrational modes. The molar heat capacity will then be the derivative with respect to temperature of the resulting sum for a one-mole specimen.

This quantum exercise, in the hands of Nernst, Einstein, and Debye, led to various expressions for the molar heat capacity of a solid. The one developed by Debye gives the best agreement with experiment:

$$c = 9R\left[4\left(\frac{kT}{hf_m}\right)^3 \int_0^{hf_m/kT} \frac{h(hf_m/kT)^3 df}{kT \exp(hf/kT) - 1} \right.$$
$$\left. - \frac{hf_m}{kT} \frac{1}{\exp(hf_m/kT) - 1} \right], \qquad (14.8)$$

where f_m is the maximum possible frequency of vibration corresponding to the situation in the solid where neighboring atoms in a row of atoms are vibrating in opposite phase.

This equation is simply bristling with expressions containing kT! Note, however, that kT does not anywhere appear alone, but always as a ratio with some energy hf or hf_m. This ratio relationship means that the interesting features of the dependence of specific heat capacity on temperature are determined by how kT compares with the vibrational energies of the solid. The Debye equation reduces to $c = 3R$ at high temperatures and goes

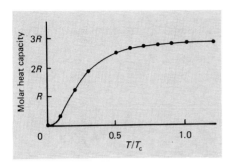

Figure 14.5 The molar specific heat of a monatomic solid is given by the Debye equation.

to zero at absolute zero temperature, in agreement with experimental observation.

Figure 14.5 is a graph of the Debye equation. The shape of this plot is identical for all monatomic solids, the only difference from material to material being in the temperatures at which the various features of the curve appear. Another inspection of equation (14.8) shows that, once the value of f_m for any material can be supplied, the entire curve can be calculated numerically. Each solid can thus be characterized by a single parameter. In practice, rather than giving f_m for a material, it is preferable to specify a related quantity T_c called the Debye characteristic temperature, such that $kT_c = hf_m$. In figure 14.5 the temperature is plotted as a fraction of the characteristic temperature T_c.

The physical interpretation of the shape of the Debye curve is as follows. At very low temperatures the thermal energy present in the specimen is too small to activate any but the lower-frequency, lower-energy modes of vibration. The higher-energy modes are dormant and do not soak up any heat energy as the temperature is raised. At intermediate temperatures, however, the higher-energy modes begin to awaken and participate in the equipartition of the input heat energy. At still higher temperatures, around the characteristic temperature, where kT is comparable with the quantum energy hf_m of the highest possible mode of vibration, the entire spectrum of modes becomes active and c is around 97% of $3R$. With further increase in temperature the molar heat capacity approaches asymptotically the classical value of $3R$, in the Dulong and Petit region.

▶ **Example 14.3** From classical statistics, what would you expect to be the molar heat capacity of a solid compound of (a) type AB and (b) type AB_2? Assume the temperature to be well above the Debye characteristic temperature.

Each atom of the molecule has three degrees of freedom, performing small vibrations about its equilibrium position. From kinetic theory, the mean kinetic energy per degree of freedom is $\frac{1}{2}kT$. In harmonic vibration, the average kinetic and potential energies are equal; the total energy per atom is $3kT$. For a molecule of n atoms the average energy is $3nkT$ and the molar heat capacity is $3nR$. So (a) for AB, $c = 6R$; (b) for AB_2, $c = 9R$.

14.6 Thermionic Emission

When a metal filament (a cathode) is enclosed in an evacuated space and heated, electrons evaporate from its surface. If a

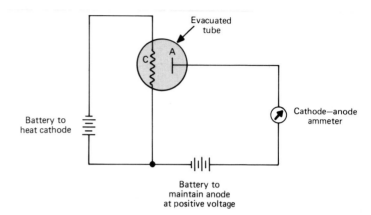

Figure 14.6 The emission current of a heated metal surface can be measured with a vacuum tube in this test circuit. A, anode; C, cathode.

positively charged electrode (an anode) is also present in the evacuated space (figure 14.6), the electrons will be attracted to it and collected by it. A continuous current of electrons will exist between the cathode and anode as long as the cathode temperature is held constant and the anode voltage is maintained positive. Such an arrangement is at the heart of the vacuum tubes formerly used in radio receivers.

This electron evaporation process is similar to the process whereby molecules of a liquid evaporate from a liquid surface. The evaporating molecules must have enough kinetic energy to tear themselves away from the surface against the restraining forces of the molecules left behind in the liquid. At low temperatures only a small fraction of the molecules in the surface layer of the liquid will have this much energy, so the evaporation will be slow. By heating the liquid one can increase this fraction and cause the evaporation rate to rise.

In a diode if the anode is sufficiently positive to collect all the electrons emitted by the cathode before something else happens to them, it is clear that the current will be determined by the rate at which the electrons evaporate. This rate depends on the temperature of the surface. The resulting current is given by Richardson's equation

$$I = AT^{1/2} \, e^{-\phi/kT}, \tag{14.9}$$

where A is a constant characteristic of the particular emitting surface and ϕ is the activation energy, that is the kinetic energy an electron must have to break away from the surface once it arrives there by random diffusion.

What insight can we obtain from the Richardson expression regarding the pivotal significance of kT? First, let's simplify this task and recognize that as the temperature of the cathode is

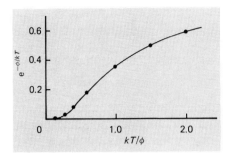

Figure 14.7 The emission current factor $e^{-\phi/kT}$ increases with increasing temperature.

varied, most of the variation of the evaporation current results from the exponential factor $e^{-\phi/kT}$ rather than from the $T^{1/2}$ factor. Let's disregard $T^{1/2}$ for the time being and make a plot of $e^{-\phi/kT}$ to see how it varies with temperature. This is shown in figure 14.7 with the abscissa giving, not the temperature explicitly, but rather the product kT as a fraction of ϕ. The plot shows that there is very little emission at low temperatures and that the emission current doesn't become substantial until the temperature is raised high enough so that kT is at least 0.2ϕ. This finding illustrates the generalization that whenever an atomic or molecular process depends on temperature, the process doesn't go very fast or very far unless the temperature is high enough to ensure that kT is a significant fraction of the activation energy for the process.

It may be difficult for you to see how thermionic emission can occur at all when $\frac{3}{2}kT$ is anything less than ϕ. The reason is that we are dealing with a statistical spread of electron kinetic energies. Just as in the case of the distribution of molecular velocities, if the bulk of the electrons is moving with thermal energy around $\frac{3}{2}kT \simeq 0.3\phi$, a small fraction of them, at the upper tail of the distribution, will have energies in excess of ϕ and will be candidates for evaporation. In order to make good their escape, of course, the favored few must not only have the required energy but must also be near enough to the surface to reach it within the next free path, and must be moving in the direction of the surface.

14.7 Electrical Conductivity in Solids

Our model of conductivity in solids pictures the ions of the solid as being permanently anchored to their crystal lattice positions (except for vibration), while the valence electrons of those ions are stripped free to percolate through the lattice, and, by their motion in an impressed electric field, to transport charge and thus produce a current. An electron in an electric field in a solid will accelerate in the direction of the field until it collides with an ion, transferring to it some of the kinetic energy the electron acquired during its acceleration and starting out afresh on another accelerating path in some other random direction, but always being pulled around towards the direction of the field. This drift motion in the field is superposed upon the random thermal motion as a statistically steady current. The percolation of the electrons through the ion structure resembles the motion of a group of marbles rolling down an inclined sheet of plywood from which many nails protrude. The marbles do not accelerate indefinitely as they roll down, but reach a statistically steady falling speed determined by the collisions they are continually making with the nails.

Solids are classified into three types according to the readiness with which they conduct current. First are the good conductors, mostly metallic in chemical nature, characterized by electrical conductivities in the range from 10^3 to 10^6 mho/cm†. Then come the semiconductors, materials such as silicon, germanium, selenium, copper oxide, zinc oxide, and lead sulfide, having conductivities in the range from 10^{-2} to 10^{-8} mho/cm. Finally, there are the insulators, characterized by conductivities less than 10^{-12} mho/cm. Any theory of electrical conduction must be capable of explaining the enormous range of conductivities exhibited by these materials.

Suppose we look first at the semiconductors. We imagine that the valence-electron population inside a semiconductor is distributed in two possible states, one in which the electrons are completely free to move as described above, the other in which, although they can move about in the lattice, there are constraints which prevent them from participating in a current. These two states are designated 'valence band' and 'conduction band' in figure 11.4. They differ in energy by the amount ΔE per electron and at ordinary temperatures the electrons are in the lower-energy, non-conducting state. Only a very small fraction of the electron population inhabits the higher-energy, conducting state. It is this small number of electrons which is responsible for the feeble conductivities these materials possess. If the temperature is increased, the probability that an electron can jump from the non-conducting to the conducting state is improved, with the result that more of them do it, thus increasing the conductivity of the material. You can be sure that the resulting conductivity is going to depend on how kT compares with the activation energy for promoting an electron into the upper, conducting state.

Theory based on this model predicts that the conductivity σ of a semiconductor should vary with temperature according to the expression

$$\sigma = \sigma_\infty \, e^{-\Delta E/2kT}, \tag{14.10}$$

where σ_∞ is the conductivity of the material at some extremely high temperature, not realizable in practice, at which the valence-electron population of the semiconductor should be equally

† Since the unit for resistance is the *ohm*, conductance, the reciprocal of resistance, is assigned the unit *mho*.

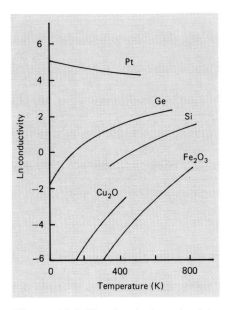

Figure 14.8 The electrical conductivity of a semiconductor increases with increasing temperature.

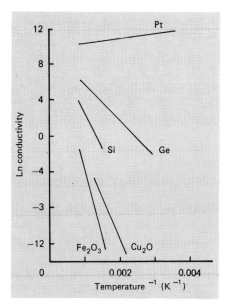

Figure 14.9 Straight-line plots are obtained when the conductivity curves of figure 14.8 are replotted on suitable axes.

divided between the two states. This equation should be compared with equation (14.9) for thermionic emission from the surface of a solid. You see the same kind of exponential dependence of the phenomenon upon a quantity involving the ratio of kT to the activation energy for the process. Indeed, the conductivity of a semiconductor may be thought of as due to a kind of thermionic emission of electrons from a non-conducting state into a conducting state in the interior of the material. Figure 14.8 shows experimental plots of ln conductivity against temperature for a few typical semiconductors. The logarithmic scale of conductivity is necessary to accommodate the large ranges of conductivity displayed.

Evident at once is the large variation of conductivity with temperature. For a typical metal around room temperature, the conductivity *decreases* by about 1/273 of its value for every degree increase in temperature. On the other hand, the conductivities of some semiconductors *increase* by as much as 5% per degree increase in temperature. This fact is made use of in sensitive temperature measurement and control using semiconductor circuit elements.

Confronted by a set of curves such as those of figure 14.8, a scientist usually attempts to replot the data on different axes in such a way as to obtain straight lines. Equation (14.10) suggests plotting the logarithms of the conductivities against the *reciprocal* of the absolute temperature. When this is done one obtains the family of curves shown in figure 14.9. The straightness of these curves shows the excellent agreement between theory and reality as to the exponential nature of the dependence of conductivity on temperature.

Taking natural logarithms in equation (14.10) we obtain

$$\ln \sigma = \ln \sigma_\infty - \Delta E / 2kT. \qquad (14.11)$$

Thus the intercept on the vertical axis of any of the lines in figure 14.9 gives the value of $\ln \sigma_\infty$ for the material, while the slope of the line gives the value of $\Delta E/2k$, from which the activation energy ΔE can be calculated.

It is significant that all the lines in figure 14.9 appear to converge to a common zero intercept around $\ln \sigma_\infty \approx 10$, even the line for the typical *metal*, platinum. We may conclude that at sufficiently high temperatures semiconductors would become essentially metallic in their conduction behavior; that is, if they didn't melt first.

Now let's look at the metals and the insulators. To answer the question why metals have conductivities so much larger than those of semiconductors at room temperature, we have only to observe that the two states either overlap each other or are separated by a negligibly small ΔE. Thus in metals a large fraction of the valence-electron population is already in the conducting state, even at room temperature. With lots of electrons to conduct, the conductivities of metals are high.

The vanishingly low conductivities of insulating materials are attributable to the comparatively large energy gap between the conducting and non-conducting states. In these materials ΔE is many times larger than kT even at the highest temperatures attainable without melting. The electrons cannot make it into the conducting state in numbers sufficient to give substantial conductivity.

14.8 Concluding Remarks

We have given enough instances to illustrate the range of physical situations which involve the product kT. We now recapitulate the substance of this chapter.

(a) The most probable kinetic energy of translation of a system of particles is kT per particle.

(b) The RMS kinetic energy of translation of a system of particles is $\frac{3}{2} kT$ per particle.

(c) The internal thermal energy of a system of particles at temperature T is $\frac{1}{2}kT$ per particle per degree of freedom for all those degrees of freedom which are fully active at that temperature.

(d) The product kT is often found in ratio with the activation energy for some processes in an expression giving the rate or extent of the process. Since so many of the actions and processes falling within the realm of physics depend on what individual atoms, molecules, ions, or electrons are doing, the product kT is bound to appear frequently.

▶ **Exercises**

14.1   There is spectroscopic evidence for the existence of much hydrogen in the Sun. Find the speed v_{RMS} of hydrogen molecules near the surface where the temperature is about 6000 K.

14.2   For what temperature does kT equal 1 eV?

14.3   In a nuclear explosion of ^{238}U the temperature at the center is of the order of 120×10^6 K, the density ρ of uranium is 20 g cm^{-3}. (a) Find the average kinetic energy of nuclei at this temperature. (b) Find the pressure at the center of the explosion before the products have started to disperse.

14.4   Estimate the mean free path l of a 2.39 inch diameter billiard ball when randomly aimed on a table 4.0 ft × 8.0 ft on which lie two other similar balls.

14.5   Find the number of collisions per second of oxygen molecules at 500 K and 1 mmHg if the diameter of an oxygen atom is 3.0×10^{-10} m.

14.6   In helium gas at 1 atm the mean free path of a molecule is 1.86 $\times 10^{-7}$ m. Show that the kinetic theory diameter of a helium molecule is about 2.5×10^{-10} m.

14.7   (a) At what height in an atmosphere at a uniform temperature of $0\,°$C is the pressure just half that at sea level? (b) What further height would again reduce the pressure by one half?

14.8   In the Earth's atmosphere, air molecules have speeds given by Maxwell's distribution law. Some will have speeds greater than $v_e = \sqrt{(2g_0 R)}$ needed for escape from Earth and hence will leave the atmosphere. This will continue until all the atmosphere is lost to outer space. Why hasn't this already occurred?

14.9   What is the ratio of the number of hydrogen molecules n_1 whose speeds lie in the range 3000–3010 m s^{-1} to the number n_2 whose speeds lie in the range 1000–1010 m s^{-1} if the temperature of the gas is 300 K?

14.10   By analogy with Richardson's equation for thermionic emission, write an expression for the rate of sublimation from a solid surface having a molar heat of sublimation H_s.

15 Noise

'Till e'en those bodies move that we perceive
In motion in the sunbeam, yet can ne'er
With sure discernment note those blows whereby
They are constrained to act in such a wise.

Lucretius
De Rerum Natura, Book II, lines 141–44

15.1 Introduction

Now that you have read and, we hope, digested the account in Chapter 14 of the everywhere-and-forever kT, it is appropriate to consider a particular consequence of the eternal restlessness of atoms, molecules, ions, and elementary particles, owing to thermal agitation. This chapter deals with the various manifestations of what is called thermal noise.

The layman thinks of noise as any sound which is unpleasant or annoying. However, the scientist defines noise as any disturbance, acoustical, mechanical, electrical, or whatever, that masks or even perceptibily interferes with the measurement he is undertaking. The thermal motions of the ultimate particles of nature present a limit to the precision with which any measurement or observation can be made.

15.2 Robert Brown's Irritable Particles—Mechanical Thermal Noise

In 1827 Robert Brown, a Scottish botanist, using an ordinary microscope with double-convex lens, noted that cytoplasmic granules from pollens when dispersed in water exhibited chaotic movements. It is likely that others looking at water through a microscope had also seen little things moving about. Brown showed that ground glass, powdered minerals, even stone dust from the Egyptian Sphinx, exhibited these motions, proving that behavior of the 'animated or irritable' particles did not depend on their being alive.

In the years that followed, convection currents, uneven evaporation, intermolecular forces, hygroscopic and capillary action, bubble formation, heating of the liquid from illumination, and the effect of electrolytes were all investigated and shown *not* to be the cause of Brownian motion. Through the rest of the nineteenth century this phenomenon absorbed the interest of chemists, physicists, and biologists, with the growing conviction that the random particle motion resulted from thermal bombardment by molecules of the suspending medium. In 1905 (the fruitful year in which he also explained photoelectricity and launched relativity) Einstein made a mathematical analysis based on the molecular bombardment model. He obtained results which were quantitatively verified by Jean Perrin in a series of experiments with particles of known uniform size suspended in liquids of known viscosity. Perrin's results led to the full acceptance of the molecular view of matter. Let's have a closer look.

Any solid particle suspended in a liquid or gas will suffer impacts with the molecules of the medium as they dart about with thermal speeds. Since the molecular motion is random, in any short time interval Δt the particle will suffer more collisions from one side than it will receive in the opposite direction from the other side. It will thus experience a net force and will respond by accelerating in the direction of that force. The acceleration is tempered by the viscous drag of the suspending medium. Since the net force of the unbalanced molecular impacts is continually changing in direction and magnitude, the resulting motion of the particle will be random, continually changing in direction and speed.

It was shown in Chapter 14 that the average translational speed of a molecule is, from equation (14.5),

$$v = \sqrt{\frac{3kT}{m}}.$$

This equation says that the thermal speed of a particle is inversely proportional to the square root of its mass. One of the fundamental postulates of the kinetic theory is that in thermal equilibrium the internal energy of a system of many particles, on average, is equally divided among all the degrees of freedom of all the particles. This principle of equipartition holds whether the particles are all alike (as in a specimen of pure water), whether they include foreign molecules of different molecular weight, or whether they include colloidal particles in suspension. Thus, a colloidal particle in thermal equilibrium with the medium it is suspended in might be expected to have a thermal speed of $\sqrt{(3kT/m)}$. For a typical colloidal particle having a mass 10^{12} times greater than that of a water molecule, this average thermal speed would be around $0.0007\,\mathrm{m\,s^{-1}}$. However, the particle does not actually move with this speed. The accelerating force due to the unbalanced buffeting of the molecules of the suspending medium is opposed by the viscous drag of the medium itself. Resulting particle speeds are typically around 1% of this value.

If the position of a spherical colloidal particle is noted and plotted for a considerable interval of time t, the particle will be found, on average, to have wandered farther and farther from the place where it was observed to start. Einstein derived an expression for this displacement as:

$$\bar{x}^2 = kT\frac{t}{3\pi r\eta}, \tag{15.1}$$

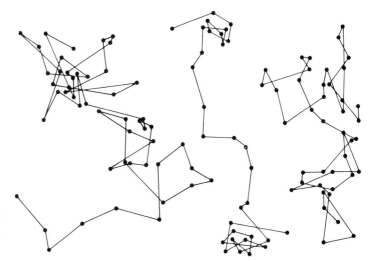

Figure 15.1 These three sequences were obtained by tracing the segments which join the consecutive positions of the same water-suspended granules of mastic at intervals of 30 s.

where $\bar{x}$ is the average magnitude of the displacement† without regard to its direction, which will be random; r is the radius of the particle, and η the coefficient of viscosity of the medium. There's that kT again! Figure 15.1 shows sketches, similar to those obtained by Perrin, with straight lines joining the successive positions of suspended particles executing Brownian motion, recorded at 30 s intervals.

▶ **Example 15.1**. A colloidal particle of radius 10^{-7} m suspended in a liquid of the same density and viscosity 0.001 N s m^{-2} is observed at two instants 100 s apart. The observations are repeated many times. On average, how far will the particle have moved during the 100 s interval, at room temperature?

From equation (15.1) we have

$$\bar{x}^2 = kT\frac{1}{3\pi r\eta}.$$

In this example $T = 300$ K, $r = 10^{-7}$ m, $t = 100$ s, $\eta = 0.001$ N s m^{-2} and $k = 1.38 \times 10^{-23}$. Therefore

$$\bar{x}^2 = \frac{1.38 \times 10^{-23} \times 300 \times 100}{3\pi \times 10^{-7} \times 0.001},$$

$$- 4.4 \times 10^{-10} \text{ m}^2.$$

† Displacement means not the total distance that the particle has traveled in all its loops, kinks, and wanderings, but rather the *straightline* distance from where it started to the place where it is observed at a subsequent time t.

so

$$\bar{x} = 2.1 \times 10^{-5} \text{ m} = 2.1 \times 10^{-3} \text{ cm}.$$

Brownian motion has been found to be troublesome when one makes very delicate current measurements with a sensitive galvanometer having a suspended-mirror-and-scale reading arrangement. The galvanometer mirror, affected by random molecular bombardment, exhibits an angular Brownian motion of its own. Also, the random thermal motions of the electrons in the wire of the galvanometer coil cause a continually fluctuating torque. The image of the scale, as viewed through the reading telescope, never settles down, and the observer must content himself with a time-average reading.

15.3 Electrical Thermal Noise—Resistance Noise

Consider a length of metal wire lying on a table top. In the metallic state the valence electrons belonging to the atoms of which the wire is made are detached from their parent atoms, and are thus free to bounce about among the metal ions, with thermal speeds and in ever-changing random directions, like the molecules of a gas in a container. In fact, the term 'electron gas' has been applied to such a valence-electron population. This electron gas fills the interior of the wire with a density which is macroscopically uniform from place to place. However, on an atomic, microscopic scale, there are continual, fleeting fluctuations from this uniform density. In a cubic volume element with ten atoms per side there may be, at some particular instant, 986 free electrons. A tenth of a microsecond later there may be 1029. Any tendency for random deviations from uniform local electron density is, of course, opposed by local space charges which immediately disperse the excesses and fill up the deficits. However, the fluctuations continue because of the ceaseless random flight of the electrons.

Imagine a voltmeter of high sensitivity to be connected across the ends of this piece of wire. We shall assume, for the moment, that this voltmeter and its connecting wires are ideally noiseless. At some particular instant there may be a few more electrons in the left-hand half of the wire than in the right-hand half. At this instant the voltmeter will indicate a voltage. In the next instant the electron population in the wire will have redistributed itself in some different configuration, so that the voltmeter will read a different voltage between the ends of the wire. This fluctuating voltage continues ceaselessly, having an average amplitude and a particular frequency distribution. Let's investigate these properties further.

First, let's look at the amplitude of the fluctuation. We shall approach this matter by enquiring about the degrees of freedom of the wire, recognizing that each degree of freedom has an associated average thermal energy of $\frac{1}{2}kT$, and summing over the total number of degrees of freedom. The number of degrees of freedom of our wire will be the number of different normal modes in which it can oscillate electrically.

Like a stretched string which can vibrate mechanically when plucked or bowed, so can a length of wire oscillate electrically when excited. The vibration or oscillation can occur in any number of different normal modes, with the lowest-frequency mode corresponding to the vibration of the entire length of the string as a whole or the oscillation of current in the entire length of the wire as a whole. In the second mode the string vibrates in two segments at twice the frequency of the first mode. Similarly, in the second mode of electrical oscillation of the wire the current exhibits two loops, with a node in the middle of the wire. The frequency of this mode is twice that of the lowest mode. And so on.

The frequencies of the modes of vibration of the string form the spectrum:

$$\sum_n f_n = \sum_n \frac{nv}{2l},$$

where f_n is the frequency of the nth mode, l is the length of the string, and v is the speed of propagation of a transverse wave along the string. Similarly, the spectrum of normal modes of electrical oscillation in the wire is given by

$$\sum_n f_n = \sum_n \frac{nv}{2l},$$

where v is the speed of propagation of a voltage wave along the wire. The problem now becomes that of finding how many of these modes are included in the frequency range to which our ideal voltmeter responds. Each of these modes exhibits an average energy of magnitude $\frac{1}{2}kT$ from the thermal agitation of the electrons which maintains the oscillations. The AC fluctuating voltage indicated by the voltmeter will be given by a summation which was first carried out by H Nyquist in 1928, with the result:

$$\bar{e}^2 = 4kTR\Delta f, \tag{15.2}$$

where $\bar{e}^2$ is the mean squared AC voltage, R is the resistance of the wire, and f is the width of the frequency band to which the voltmeter is responsive. In the same year, J B Johnson experimentally verified the dependence of the thermal-noise voltage on the resistance of the element, be it a metal wire such as we have been considering, a carbon filament, or a water solution of an electrolyte. The thermal noise between the terminals of a resistor is often called *Johnson noise*. The formula above holds throughout the audio, radio, and microwave frequency ranges, but it requires modification for frequencies so high that the quantum energy hf of the oscillations becomes comparable with kT. The complete expression, valid for all frequencies, is obtained by replacing kT in equation (15.2) by the quantity $hf/(e^{hf/kT}-1)$, obtaining:

$$\bar{e}^2 = \frac{4Rhf}{e^{(hf/kT)}-1}\Delta f. \qquad (15.3)$$

To orient you as to magnitudes, the thermal noise voltage of a $1000\,\Omega$ resistor, as measured by an ideal AC voltmeter having a frequency response band 1 MHz wide, is about $4\,\mu$V at room temperature.

We have been speaking of ideal voltmeters. Of course, there is no such thing. All voltmeters have internal resistance (the higher the better), and any attempt to measure electrical thermal-noise voltages with a real voltmeter will be complicated by the thermal noise of the resistance of the instrument itself!

▶ **Example 15.2** In an attempt to measure the noise voltage of a 1 MΩ resistance R, an experimenter connects it to an AC voltmeter having internal resistance $R_v = 20\,$KΩ. What does he read?

From equation (15.2), the noise EMF of the resistance R is $\sqrt{(4kTR\Delta f)}$, so the circuit current will be $\sqrt{(4kTR\Delta f)}/(R+R_v)$, and the voltage indicated by the meter will be $R_v\sqrt{(4kTR\Delta f)}/(R+R_v)$. Hence the voltmeter reading will be $R_v/(R+R_v)$ times the open circuit noise voltage of R. Here $R_v = 20\,$kΩ and $R = 1\,$MΩ, so $R_v/(R+R_v)$ is $20\,000/1\,020\,000 = 1/51$ and the voltmeter will record a voltage of $1/51$ times the noise EMF of the resistance. Only if the voltmeter resistance is many times greater than R will its reading be anywhere near the true noise EMF of R.

Thermal noise presents a severe limiting problem in the reception of very weak radio signals. Just as stars cannot be seen against the interfering brightness of the sky in the daytime, it is easy to understand that a radio signal which is weaker than the ever-present noise may not be detected. However, the sensitivity of radio receivers can be increased in two ways, suggested by equation (15.2). The first is to reduce the frequency bandwidth of the receiver until is it no wider than necessary to accommodate the expected signal. This tactic excludes the noise coming from those regions of the radio spectrum which contribute nothing to the signal strength. Another procedure is to refrigerate the circuit element in the receiver which is chiefly responsible for the noise. This is usually the grid resistance or the emitter-circuit resistance of the first stage of the receiver amplifier.

15.4 Acoustical Thermal Noise

Obviously, acoustical thermal noise is forever present. The random fluctuations of air pressure on the ear drum, even in the absence of any sound-producing agency except molecular agitation itself, cause ceaseless vibrations of the ear drum over a wide spectrum of frequencies. These vibrations are just below the threshold of hearing and cause no annoyance. If the human ear were only an order of magnitude more sensitive, however, a faint rushing or rustling sound would be present as a background to everything we hear.

It is tantalizing to recognize that no measurement of acoustic thermal noise can ever be made with equipment at the same temperature as the air. What could you measure noise with that would not introduce thermal noise of its own into the measurement?

15.5 Optical Thermal Noise—Blackbody Radiation

The thermal agitation of atoms and molecules produces thermal noise in optical systems, as well as in mechanical, electrical, and acoustical systems. Every surface radiates and absorbs energy at a rate appropriate to its temperature. Imagine a closed room with walls, ceiling, and floor at temperature T. The room is filled with radiation which, in thermal equilibrium with the surfaces exposed, has an energy density and spectral distribution given by the classical Rayleigh–Jeans equation as

$$\rho = \frac{8\pi f^2 kT \Delta f}{c^3}, \qquad (15.4)$$

where ρ is the radiant energy per unit volume in the frequency range Δf, h is Planck's constant, f is the central frequency of the interval Δf, and c is the speed of light. The spectral distribution of this thermal-radiation noise is plotted as the left-hand straight portions of the curves shown in figure 15.2.

This equation gives good agreement with experiment at frequencies up to those in the infrared region of the spectrum. However, it is worthless for frequencies in the visible region and higher, for which the quantum energy hf of the radiation becomes comparable with or greater than kT. Equation (15.4) predicts an ever-increasing energy density with increasing frequency. Obviously, since the energy density cannot approach infinity as the bandwidth is increased, some modification of this equation is required to adapt it for high frequencies. Such a modification was introduced in 1900 by Max Planck, who replaced kT by the quantity $hf/(e^{hf/kT} - 1)$, to obtain:

$$\rho = \frac{8\pi^2 hf^3}{c^3 (e^{hf/kT} - 1)} \Delta f. \tag{15.5}$$

This equation is known as the blackbody distribution law. The quantum concept on which Planck's derivation was based caused classical physics to evolve into 'modern' physics during the early decades of the twentieth century. The right-hand portions of the curves of figure 15.2 were calculated from Planck's equation.

Any optical measurement must be made in the presence of this

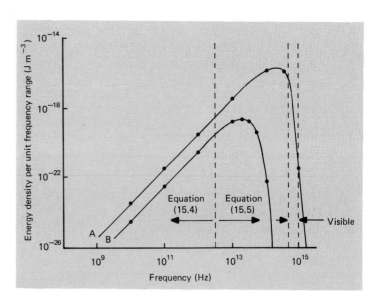

Figure 15.2 The radiant energy density of a blackbody at two different temperatures is shown as calculated from the classical Rayleigh–Jeans law for low frequencies and from the Planck law for higher frequencies. Curve A, 3000 K; Curve B, 300 K.

background radiation from the surroundings. If a room were to be totally darkened, and if the room and its contents were all at the same temperature, it would be impossible to detect any object in the room by its own or by reflected radiation. Only if the object were warmer or colder than its surroundings could it be picked out by any optical device.

15.6 Thermal Noise—Conclusion

We see, then, that thermal noise plays a role in many realms of physics, in particular those which deal with the ultimate particles of nature. In the chapters of this book we have endeavored to develop various phenomena in ways which permit some master formula to describe a phenomenon in its several appearances in mechanics, acoustics, electricity, heat, etc. The phenomenon of thermal noise, however, we have chosen to present much as it has evolved classically. It should be no surprise that all of the expressions reproduced here involve the product kT. The principal point to be made is that thermal noise is an all-pervading phenomenon which places a limit on the ultimate sensitivity with which any measurement can be made, whatever its nature.

15.7 Other Types of Noise

We should not wish to leave you with the belief that thermal noise is the only kind of naturally-occurring ultimate noise. Let us consider another type of noise which is inherent in the granularity of charge in electrical systems and the granularity of radiant energy in optical systems.

Suppose that water existed only in volumes of exact multiples of one cubic centimeter. Then water being delivered through a pipe would emerge from the end not in a continuous, infinitely-divisible stream, but rather in dollops of exactly one cubic centimeter each, no more, no less. The emerging stream would exhibit some random properties. For example, if the flow were so adjusted that the long-time average were $1\,\text{litre}\,\text{s}^{-1}$, it might be found that the delivery during some particular second was $997\,\text{cm}^3$. During the next second it might be $1002\,\text{cm}^3$, and so on. There would be a continuing random fluctuation of the quantity of water delivered during successive seconds because of the $1\,\text{cm}^3$ quantization of the water.

Electric charge exists in quantized bundles of exactly one electron charge each. Consider the electric circuit of figure 15.3(*a*),

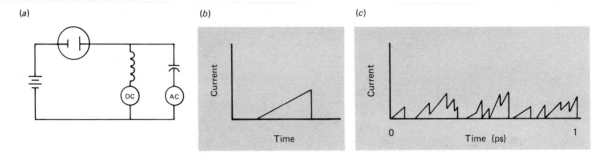

(a)   *(b)*   *(c)*

Figure 15.3 When a single electron crosses the diode in the circuit *(a)*, it causes a spike of current *(b)* in the circuit. *(c)* A 1 μA direct current shows microscopic fluctuations due to the granular nature of electric charge.

consisting of a battery, a vacuum diode, an ideal noiseless DC ammeter, and an ideal noiseless AC ammeter. Every electron which jumps across the diode gap causes a sawtooth spike of current in the circuit as it accelerates from the cathode to the anode (figure 15.3*b*). The area under this spike represents the charge of the electron.

Suppose the direct current is adjusted to 1 μA, corresponding to the flow of about 6×10^{12} electrons per second. These electrons arrive at random during that one second, and if the arrival rate could be examined microsecond by microsecond, it would show considerable fluctuation from one microsecond to the next. This fluctuation would register as an alternating current superposed upon the DC of 1 μA. The amplitude of the alternating current can be shown to be

$$\overline{i^2} = 2eI\Delta f, \qquad (15.6)$$

where e is the charge on an electron, I is the long-time average direct current, and Δf is the frequency range over which the AC fluctuation is measured. The derivation of this expression assumes that the density of electrons in the gap at any instant is small enough that the flight of any electron is not affected by the charge of other electrons likewise in transit. The formula is valid for frequencies below that corresponding to the transit time of an electron across the gap.

The amplitude of this AC component for a diode carrying a current of 1 A DC is about $\frac{1}{2}$ μA AC, as measured by an AC galvanometer sensitive to frequencies from 0 to 1 MHz. This kind of noise is called *shot noise*. Note that kT is not a factor of equation (15.6). Since it is the granularity of electric charge (rather than thermal agitation) that is the agency responsible for shot noise, it is not surprising that the quantity kT should be absent.

When shot noise is amplified and heard in earphones, the sound is similar to that of rain on the roof. In the latter case one hears a roar of sustained average amplitude and frequency distribution, but the roar is made up of thousands of separate impacts per second, each one like all the others and coming at random.

Since light arrives on a surface as a bombardment by individual photons, it should not be surprising that light, even 'steady' light, exhibits a statistical randomness of intensity. If the bombarded surface is the sensitive element of a diode photoelectric cell, for example, the flow of ejected electrons will be random in time. The photocell current will therefore have an AC noise component superposed on the steady direct current. This noise is describable by equation (15.6). The photoelectric diode differs from the thermionic diode only in the mechanism by which electrons are released to start across the diode gap.

▶ **Exercises**

15.1 A 1 MΩ resistance enclosed in a thermally insulated chamber is connected in an otherwise resistanceless circuit to a 1 Ω resistance outside the chamber. The larger resistance generates a higher noise EMF than the other. Why, then, does it not feed its noise energy into the 1 Ω resistance, thereby cooling itself and turning the chamber into a refrigerator?

15.2 Note, referring to equation (15.6), that as the DC diode current increases, the shot noise fluctuation increases also. Wouldn't you expect just the opposite, since the larger direct current would give more opportunity for smoothing out the fluctuations?

15.3 The electrostatic repulsion from each other of the electrons in the electron gas inside a metal is, on average, neutralized by the presence of the positive ions. However, there are some 10^{23} electrons/cm^3 in this gas. Any sample of an ordinary gas at this particle density would exert enormous pressure. Why isn't the metal specimen blown apart by this pressure?

15.4 Calculate the transit time of an electron across a 1 cm diode gap at 100 V, and determine how many electrons, on average, are in the gap at any instant for a 100 mA diode current.

15.5 By graphical integration of the lower curve of figure 15.2, show that the approximate total energy density of radiation in a room at 300 K is 6×10^{-6} J m^{-3}.

15.6 From the curves of figure 15.2, verify that the product of the absolute temperature and the wavelength of maximum energy density is a constant. This relationship is known as Wien's displacement law.

15.7   Determine the constant in Exercise 15.6 and apply it to calculate the surface temperature of the Sun for which the wavelength of maximum emission is 5×10^{-7} m.

15.8   A radio signal of 1 μV strength is delivered to the input of a radio receiver having an input resistance of 1 kΩ. The receiver has a bandwidth of 100 kHz. Is the signal detectable in the presence of receiver noise?

15.9   A pollen grain having a density of $1.00 \, \text{g cm}^{-3}$ and radius 0.01 mm is suspended in water (viscosity $10^{-3} \, \text{N s m}^{-2}$) and observed with a microscope at room temperature. How long might you expect it to take for the grain to migrate a distance of 0.1 mm by Brownian motion?

16   Radiation Pressure

The main goal of physics is to describe a maximum of phenomena with a minimum of variables.

CERN Courier

16.1 Introduction

Have you ever felt that you had to push against sunlight as you might lean to resist a strong wind? Of course not. Yet according to Maxwell's epoch-making electromagnetic theory of light, light carries energy and momentum: when light strikes an object it exerts a force on that object, but this force is exceedingly small. It is equal to εE^2 per unit area when a plane electromagnetic wave is absorbed by a flat surface, ε being the dielectric constant of the medium and E the amplitude of the electric field. If E is a few volts per meter, εE^2 is about 10^{-9} N m^{-2}. (For comparison, in ordinary conversation the ear detects pressure changes due to the sound wave of the order of 4×10^{-2} N m^{-2}).

In 1873 Sir William Crookes thought that he had demonstrated radiation pressure in a partly evacuated bulb, only to find that the radiometer he had invented was responding to forces of molecular bombardment rather than to radiation pressure. The little rotating-vane toy often seen whirling in sunlight is *not* responding to radiation pressure; rather the torque is due to the fact that gas molecules rebound with more momentum change from the blackened (warm) surface of each vane than from the other polished (cool) surface. These radiometric forces are four or five orders of magnitude greater than the force due to radiation pressure and must be eliminated or corrected for if true radiation pressure is to be detected.

During 1902–3, Nichols and Hull succeeded in measuring the radiation pressure on one of two mirrors suspended from a torsion fiber. It was a satisfaction to have obtained this verification of electromagnetic theory, but Poynting expressed the commonly held view that the minuteness of these light forces 'appeared to put them beyond consideration in terrestrial affairs.' Invention of the laser in 1960, however, has provided light sources capable of exerting radiation forces on tiny particles strong enough to produce accelerations up to a million g. Interesting practical applications of radiation pressure now appear possible.

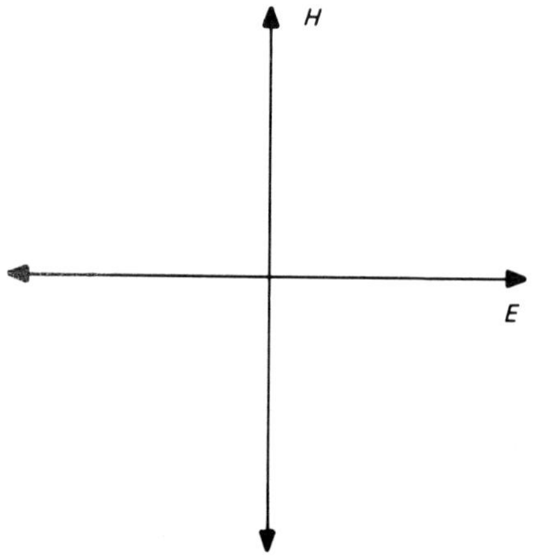

16.2 Electromagnetic Radiation Pressure

What is the physics of the pressure of electromagnetic radiation on an absorbing or reflecting surface? In a beam of radiation traveling away from you, the oscillating electric and magnetic fields (figure 16.1) are perpendicular to each other and to the direction of propagation. When this beam falls on an absorbing

Figure 16.1 The electric and magnetic fields of an electromagnetic wave are perpendicular to each other.

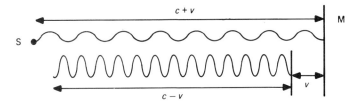

Figure 16.2 The mirror M moving forward to the left compresses the reflected wave train.

surface the electrons in the surface layers are accelerated by the electric field. The resulting motion of the electrons takes place in the magnetic field of the wave, with the result that the electron paths are curved towards the direction of propagation. At their next collision the electrons deliver to the structure of the absorber the forward momentum component thus obtained, exerting a pressure on it. If a perfect absorber is replaced by a perfect reflector, the momentum imparted is doubled and the pressure will be twice that experienced by the absorber.

The magnitude of this pressure can be obtained from Maxwell's electromagnetic field equations, but a simpler derivation, due to G F Hull,† is as follows. Imagine a train of waves of unit cross section and length $(c+v)\Delta t$ falling normally on a reflecting mirror which is moving to meet the waves with velocity v (figure 16.2). In time $\Delta t = 1$ s, all of these waves will have struck the mirror, and the first wave will be reflected back a distance c from the initial position of the mirror, or a distance $c-v$ from the mirror's position at the end of the second. The energy content of the original wave train will have been compressed from a volume $c+v$ to a smaller volume $c-v$. If this energy content is E, the energy density has been increased from $E/(c+v)$ to $E/(c-v)$. There is thus a gain of energy density of

$$E\left[\left(\frac{1}{c-v}\right)-\left(\frac{1}{c+v}\right)\right]$$

and a gain in total energy of

$$E(c+v)\left[\left(\frac{1}{c-v}\right)-\left(\frac{1}{c+v}\right)\right].$$

† Henry G E 1957 Radiation pressure *Scientific American* **196** 99–108
Lebedev P 1900 Pressure of light *International Physical Congress, Paris*, Report **2** 133–40
Nichols E F and Hull G F 1903 Pressure due to radiation *Am. Acad. Sci. Proc.* **38**(20) 559–99
Zernov X 1906 Absolute measures of sound intensity *Ann. Phys.* **21**(1) 131–40

This gain in energy must have come from the work done in advancing the mirror against radiation pressure P_r. Therefore,

$$E(c+v)\left[\left(\frac{1}{c-v}\right)-\left(\frac{1}{c+v}\right)\right] = P_r v, \qquad (16.1)$$

from which

$$P_r = \frac{2E}{c}, \qquad (16.2)$$

since $v \ll c$.

Because E/c is effectively equal to the energy density u of the incident wave train, the radiation pressure is

$$\boxed{P_r = 2u.} \qquad (16.3)$$

That is, the radiation pressure on a *reflecting* surface is simply equal to twice the energy density in the impinging beam. For an *absorbing* surface the pressure is half this value, or

$$\boxed{P_r = u.} \qquad (16.4)$$

▶ **Example 16.1** Calculate the force due to sunlight falling perpendicularly on 1 m^2 of the Earth's surface. Assume complete absorption. The solar constant σ (the rate at which radiant energy falls on 1 m^2 at normal incidence, outside the Earth's atmosphere, at the mean distance of the Earth from the Sun) is 1350 J m^{-2} s^{-1}.

From equation (16.4), the radiation pressure P_r equals the radiation density u in the impinging beam, for a perfect absorber.

$$u = \frac{\sigma}{c} = \frac{1350}{3 \times 10^8} = 4.5 \times 10^{-6}.$$

Therefore, $P_r = 4.5 \times 10^{-6} \, \mathrm{N\,m^{-2}}$.

Classical electromagnetic theory gives the energy density, equally divided between electric (E) and magnetic (H) fields, as

$$u = \tfrac{1}{2}\varepsilon_0 E^2 + \tfrac{1}{2}u_0 H^2.$$

Therefore the expression for radiation pressure P_r on a reflecting surface can be written alternatively as

$$\boxed{P_r = \varepsilon_0 E^2.} \qquad (16.5)$$

An even simpler derivation for electromagnetic radiation pressure is based on quantum concepts. A photon of light carries a quantum of energy hf, where h is Planck's constant and f is the frequency. According to Einstein's equation relating mass and energy, the photon has an equivalent mass of hf/c^2 and a momentum of hf/c, where c is the velocity of light. The total momentum contained within a beam of cross sectional area 1 m^2 and length c is Nhf, where N is the number of such photons per unit volume. This is the forward momentum delivered per second to a surface absorbing the beam. Since momentum transferred per second equals the pressure exerted, the pressure P_r is Nhf, which is the same as the energy density in the beam.

16.3 Measurement of Light Pressure

The measurement of radiation pressure, free of disturbing thermal effects, was accomplished by P N Lebedev in Russia (1901) and by E F Nichols and G F Hull in the USA during 1902–3. The latter verified with an accuracy of about 6% Maxwell's prediction that the pressure should equal the energy density of the light.

The torsion balance used by Nichols and Hull is shown in diagrammatic form in figure 16.3. Two thin disks of cover glass, C and D, each brightly silvered on one face, are suspended from a fine horizontal glass rod. The rod is suspended in a bell jar by a very fine quartz fiber, ab. The torque needed to turn the balance through a definite angle was determined from its period of free oscillation. Then from the deflection when light was thrown on one disk, one could calculate the force and from it the radiation pressure. The experimenters made their measurements at successively better (lower) vacuum pressures and corrected for the effect of residual gas atoms and the slight absorption of the silver.

16.4 Radiation Pressure of Mechanical Waves

As you may suspect, the pressure which electromagnetic (light) waves exert on a surface has a counterpart for other types of wave. For examples of mechanical radiation pressure we have only to observe that objects floating in the ocean offshore are brought in by the waves to be cast up on the beach. Also, a curtain ring threaded onto a stretched clothes'-line will move along the clothes'-line if the line is carrying a train of waves in one direction.

Let's examine the physics of the floating object with the aid of

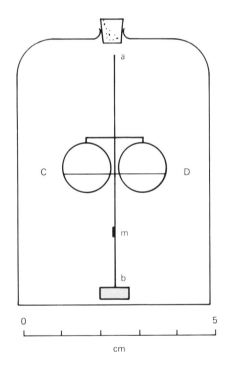

Figure 16.3 Nichols and Hull used two glass vanes, C and D, suspended on a fiber, ab, in their experiment to measure the pressure of radiation. Rotation was detected by a light beam reflected from the mirror, m.

the upper sketch in figure 16.4. Here we picture eight possible phase positions of the object with respect to the waves passing underneath it and propagating towards the right. In one complete wave period the object assumes successively all these positions, from A to A′. The object is being accelerated *upward* when it is at positions D, E and F; it is being accelerated *downward* when it is in positions H, A, and B. These accelerations, and the forces producing them, are represented by the vector arrows D, E, F; H, A, B. The vectors **B** and **F** have components urging the floating object in the direction the waves are traveling. However, the vectors **D** and **H**, of magnitudes equal to the magnitudes of **B** and **F**, urge the object with equal and opposite forces in the opposite direction. How, then, is any net horizontal motion imparted to the object over a complete wave period?

In the description above we have neglected the fact that the floating object is an oscillator, maintained in forced vertical oscillations by the waves passing beneath it. Because of the accelerations it experiences, it floats high in the water on the crest of the wave; it floats low in the water in the trough of the wave, and like all forced oscillators (see §6.5) its motion lags behind the motion of the forcing agency, in this case the water surface. It floats highest in the water not at the wave crest at A and A′, but rather at B. It floats lowest in the water not at the bottom of the trough at E, but rather at F.

With these facts in mind, we see that the top sketch of figure 16.4 gives a false picture of the situation. If we redraw the sketch to portray the buoyant forces on the object, we obtain the set of buoyancy vectors of the bottom sketch. Because of the phase lag between the motion of the object and the motion of the water surface, the buoyancy forces along the rising front of the wave, at F, G, and H, with components urging the object to the right, are larger than those along the rear slope of the wave,

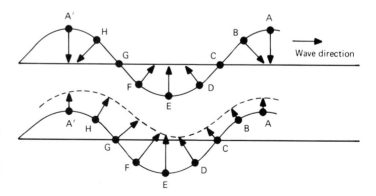

Figure 16.4 (*Top*) A floating object experiences accelerations represented by these vectors. (*Bottom*) The object experiences buoyancy forces represented by these vectors.

at B, C, and D, urging it in the other direction. The object thus experiences, during the complete wave period, a net force in the direction of the wave travel.

This argument does not give a quantitative measure of the average force acting horizontally on the floating object. However, Larmor† and Lord Rayleigh‡ concurrently analyzed the case of a train of waves traveling along a string and exerting a longitudinal force on a ring threaded onto the string. Using different approaches they both showed that the force on the ring is equal to the total (incident and reflected) linear density of the vibrational energy of the waves in front of the ring. The analogy with electromagnetic radiation pressure is thus complete.

16.5 Radiation Pressure of Sound Waves

Here we will find an analogy, but not an exact one. When electromagnetic radiation is propagated in a vacuum, the wave speed is independent of both amplitude and frequency. This is not true when the waves are traveling in a dispersive medium, nor is it true for sound waves traveling in air. Wave speed depends somewhat on amplitude; consequently there is a tendency of waves to alter their form as they proceed.

For the pressure due to sound waves, Rayleigh deduced the relation

$$P_r = \tfrac{1}{2}(\gamma + 1)E,$$ (16.6)

where E is the total volume energy density in front of the surface, P_r is the mean pressure, and γ is the ratio of specific heat capacities (see §14.3). He obtained this relation by first deriving a general expression for the mean pressure and volume of any given mass of the medium during passage of the wave. When the adiabatic relation $pv^\gamma = $ constant is inserted, the expression for P becomes that given above. If, on the other hand, we assume Boyle's law, $pv = $ constant, we must substitute 1 for γ in the expression for P which gives

$$P = E.$$ (16.7)

† Larmor J 1902 *Encyclopaedia Britannica* 10th edn vol **32** p 121
‡ Rayleigh, 3rd Baron 1902 *Phil. Mag.* **III** 338–46. Reprinted 1912 in *Scientific Papers by John William Strutt, Baron Rayleigh* (Cambridge: Cambridge University Press)

Rayleigh mentioned† that the discrepancy between $P = \frac{1}{2}(\gamma + 1)E$ and $P = E$ is related to the same causes as the tendency of waves to alter their form as they proceed.

To determine which of these relations was more nearly correct, Zernov performed measurements of the radiation pressure of sound waves. Using waves of frequency 512 Hz and very high energy density around 0.05 J m^{-3}, he found that $P = \frac{1}{2}(\gamma + 1)E$ gave results within 3% of the pressures he measured with his vibration manometer.

16.6 Consequences of Radiation Pressure

Although radiation pressure is usually very small in magnitude when compared with the other macroscopic forces acting upon bodies, there are some situations in nature where radiation pressure is the dominant agency in producing the behavior observed. We shall now proceed to examine some of these situations.

(i) Comet Tails   A comet is a body moving in orbit under the attraction of the Sun, usually consisting of a hazy gaseous cloud, a brighter nucleus, and a fainter tail. The tail is believed to consist of gas molecules and small particles evaporated from the nucleus by the heat of the Sun. Comet tails point *away* from the Sun, at least approximately, and the material in the tail is confined to the plane of the orbit: the tail is two-dimensional. It was probably Johannes Kepler who first suggested that light could exert pressure when he postulated in 1619 that it is the pressure of sunlight on the small particles that orients the tail. Although Kepler's assumption is adequate to explain some comet tails (those which astronomers call Type I, characterized by long thin streamers) it is likely that corpuscular radiation from the Sun, mostly electrons, contributes also to the strong repulsive force on the tail.

(ii) Interplanetary Dust   Consider a particle of interplanetary dust revolving in a stable circular orbit around the Sun. The equation of motion of the particle must take into account the gravitational attraction of the Sun and all the planets, the radial pressure of the sunlight, and the reaction pressure of the radiation re-emitted by the particle in all directions. On first thought one might expect that the Sun's outward radiation pressure would

† Rayleigh, 3rd Baron 1905 *Phil. Mag.* **10** 364

cause the particle to move in an orbit spiraling ever outward. Surprisingly, the particle spirals *inward*!

J H Poynting recognized (1920) that the explanation must be sought in the absorption and re-emission process. The particle absorbs light and heat from the Sun and re-radiates photons uniformly in all directions. However, as seen by an observer in a frame of reference stationary with respect to the Sun, the photons emitted by the particle in the forward direction of its orbital motion will be Doppler-shifted towards higher frequencies than those emitted in the backward direction. The energy density ahead of the particle will accordingly be greater than that behind it, and the particle experiences a net tangential radiation pressure which acts as a drag on its orbital motion. So it spirals inward and is eventually engulfed by the Sun. Most of the interplanetary dust originally present at the creation of the solar system has long ago been swept up by this process.

(iii) Quasi Stellar Objects Under some conditions, radiation pressure is believed to provide the dominant force on gas or dust clouds near quasi-stellar objects (QSOs). The gravitational force on a (spherical) particle is proportional to its mass, that is to the cube of its radius, but radiation pressure is proportional to the surface area, that is to the square of the radius. Such a particle, released at rest in the vicinity of a central source of gravitation and radiation, will be repelled if it is smaller than a certain critical size, attracted if it is larger. Since, in Newton's theory, radiation pressure and gravitational attraction both vary inversely as the square of the distance to the central source, the net repulsion or attraction will persist at all distances and the particle will be either permanently driven away or swallowed up.

Applying general relativity theory, P D Nordlinger deduced that the force of radiation pressure should decrease more rapidly than the force of gravitation with increasing distance from the source; thus a particle initially repelled by radiation pressure will retreat outward to such a distance that the radiational repulsion is just balanced by the gravitational attraction. It will then remain permanently at this equilibrium distance and will return thereto if displaced.

(iv) Satellite Orbits For an Earth satellite revolving in an orbit lying entirely within the sunlight, or in a perfectly circular orbit, the net acceleration a_p due to radiation pressure is zero. For half the orbit a_p is positive, for the other half it is negative, symmetrically. But for an elliptic orbit which passes through the Earth's shadow there is a net change in total orbit energy of the satellite.

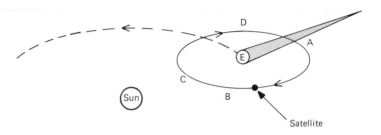

Figure 16.5 An Earth satellite experiences no acceleration from radiation pressure while orbiting through the Earth's shadow.

Consider the orbit situation depicted in figure 16.5. While the satellite is moving over the portion ABC of the orbit it experiences a retardation due to the radiation pressure of the sunlight. On the orbit portion CDA it experiences a compensating acceleration, but the compensation is not complete because part of this orbit segment lies in the Earth's shadow. As a result the satellite loses a small amount of energy on each revolution, so the perigee altitude decreases.

For Echo-type balloon satellites, having large surface and small mass, the effect of radiation pressure was sufficient to affect the perigee altitude enough to measure. Good agreement was found between the measured and expected altitudes.

(v)Solar Sail Spacecraft Proposals for solar propulsion of spacecraft have been of two categories. In one, solar rays are used to heat hydrogen gas. The gas is expelled through a nozzle to produce thrust. In the other category are schemes to have a large, light-weight surface attached to the spacecraft (figure 16.6) on which the radiation pressure of solar rays would provide a thrust. Once a spacecraft is well removed from the Earth or other planet, or is established in a satellite orbit, even a small thrust will alter its path or accelerate its flight significantly. It might conceivably bring the speed of the spacecraft v sufficiently close to the speed of light c so that the time dilation relation of relativity, $\Delta t = \Delta t_0 / \sqrt{(1 - v^2/c^2)}$, would allow an astronaut to reach the vicinity of another star, say Altair (16 light-years distant), during his lifetime. For the return journey the astronaut would orient his sail to receive thrust from the radiation of Altair. After returning to the point where solar radiation is equal to Altair's radiation, he would feather his sail and coast the rest of the way home down the Sun's gravity gradient, using radiation pressure once more to check his momentum when he got close to home.

(vi) Levitation by Laser A laser source can provide essentially monochromatic coherent light with energy concentrated in a very narrow beam in which the light intensity is some 10 000 times greater than that in the entire visible spectrum at the surface of

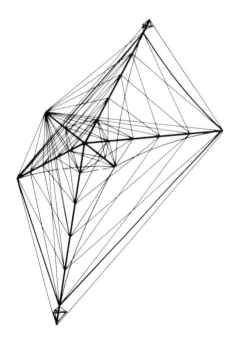

Figure 16.6 A contemplated version of a solar-sail spacecraft.

the Sun. The pressure of this radiation is readily detectable. It may serve to trap and manipulate drops and other particles in ways helpful to studies in cloud physics, aerosol science, fluid dynamics, and optics. It makes it possible to observe interactions of the drops with an electric field, the surrounding gas, and with other drops. Practical applications seem promising in what was an esoteric field of physics at the time the first measurements of radiation pressure were made by Lebedev and by Nichols and Hull.

▶ **Example 16.2** (a) Estimate the force exerted on a 100% reflecting metallic-coated spherical reflecting drop of $0.5\,\mu$m diameter by 1 W of continuous green ($0.5\,\mu$m) laser light focused on the drop. (b) What acceleration would this force produce?

(a) $1\,\text{W} = 1\,\text{J s}^{-1}$. There is 1 J in a cylindrical volume of cross sectional area πr^2 and length c. The energy density u of incident radiation is $1/\pi r^2 c$. (Since we are considering a spherical surface, we do not need the factor of 2 that we should use for reflection from a plane surface.)

$$\text{Radiation pressure } P_r = \frac{1}{\pi r^2 c}.$$

$$\text{Force } f = P_r A,$$

$$= \frac{\pi r^2}{\pi r^2 c} = \frac{1}{c}.$$

Therefore force on drop $f = 3.3 \times 10^{-9}\,\text{N}$.
(b) Assume density ρ of drop $= 1\,\text{g cm}^{-3} = 10^3\,\text{kg m}^{-3}$. Then mass $= \rho \frac{4}{3}\pi r^3 = 6.5 \times 10^{-17}\,\text{kg}$. Acceleration $a = f/m$, so

$$a = \frac{3.3 \times 10^{-9}}{6.5 \times 10^{-17}},$$

$$= 5.1 \times 10^7\,\text{m s}^{-2}.$$

So the acceleration is just over 5 million gs!

In experiments described by A Ashkin,† transparent plastic spheres of two sizes (2.5 and $0.5\,\mu$m) were suspended in neutral equilibrium in a water-filled cell and were illuminated by a horizontal laser beam. The magnitude of the force on a plastic sphere can be found by adding the effects of all the light rays striking it. Each ray is partially reflected and refracted at the surface of the sphere and contributes to the net force. From the value of the force when the particle is on the axis of the beam and from Stokes's law, one can predict the velocity of the sphere

† Ashkin A 1972 The pressure of laser light *Sci. Am.* **226** 63–71

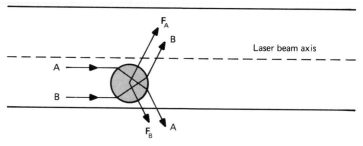

Figure 16.7 A particle in a laser beam experiences a transverse force.

through the water in the direction of the beam, larger particles having higher velocities. These predictions agree with measured velocities and indicate that the radiation force is the only significant force operative.

When the beam is turned off momentarily, a sphere originally on the axis wanders off randomly, but it returns to the beam center when the beam is turned on again. The cause of this interesting focusing action is shown in figure 16.7 to depend on the index of refraction of the sphere relative to its surroundings. The diagram shows the forces associated with the refraction of a pair of typical light rays (A and B) as they pass through an off-axis sphere that has a higher index of refraction than its surroundings. Force F_A is greater than force F_B; there is a net transverse component of force impelling the sphere back toward the axis of the beam.

Figure 16.8 shows experimental apparatus for using radiation pressure to levitate liquid drops in air against the force of gravity. The laser beam is aligned with a 0.5 mm hole H located in a sliding roof cover C. An atomizer A sprays a cloud of liquid droplets into storage vessel V. Some droplets as they settle fall through the hole and enter the light beam where, if their sizes are in the correct range, they can be trapped and levitated. An enlarged view from microscope M_1 is projected onto a screen for ease of observation, measurement and control. An electric field can be applied to the drops by a potential between plates E_1 and E_2. Once drops have been collected by the beam, the opening H can be covered by the sliding glass plate G, the vessel V removed, and microscope M_2 used to view the particles from above. The apparatus can be used to measure charges carried by drops, rates of evaporation, condensation, charging and neutralization.

Another use for optical levitation is the study of light scattering from single particles. Figure 16.9 displays the scattered light distribution from an approximately 20 μm diameter glass particle on a card held above the sphere. This type of scattering is called Mie scattering. It is widely used in the measurement of sizes for particles which are large compared with a wavelength.

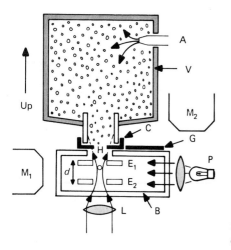

Figure 16.8 A liquid droplet is suspended by the radiation pressure of a laser beam. Dimension *d* is about 0.6 cm.

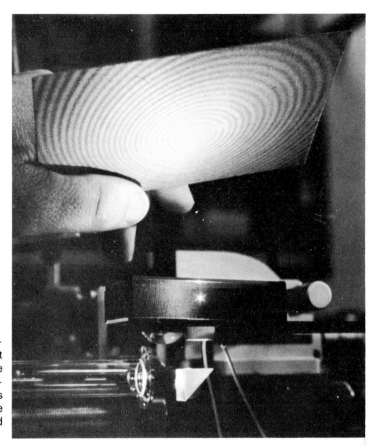

Figure 16.9 This example of Mie scattering shows the distribution of light scattered from a 20μm glass particle. The glass particle, appearing as the six-pointed star in the center of the picture, is levitated by the radiation pressure of the light beam. (Courtesy of Dr A Ashkin and Bell Laboratories.)

▶ **Exercises**

16.1   When the Earth satellite of figure 16.5 loses energy because of interrupted solar radiation pressure, what becomes of the lost energy?

16.2   Sketch an orbit situation similar to that of figure 16.5 so that the satellite *gains* some net energy in each revolution.

16.3   Why does the satellite of figure 16.5 neither gain nor lose energy because of radiation pressure if it is in a *circular* orbit?

16.4   By considering a diagram similar to that of figure 16.7, show that a sphere will be forced laterally out of the beam if the index of refraction of the sphere is less than that of its surroundings.

16.5   Why, do you suppose, were transparent spheres chosen for the radiation pressure experiments of figure 16.7 rather than spheres with highly reflecting metal surfaces or spheres with black surfaces?

16.6   Compare the apparatus shown in figure 16.8 with the Millikan oil-drop apparatus, as to purpose and operation.

16.7   Using the result of Example 16.1, show that the total force of solar radiation pressure on the Earth is about 750 MN. Assume average reflectivity of 25% and the radius of the Earth to be 6500 km.

16.8   In space, a reflecting particle is attracted to the Sun by gravity and repelled by its radiation. If the density of a particle is 2.0 g cm^{-3}, find its diameter for the two forces to balance when the particle is one Sun–Earth distance (150×10^6 km) from the Sun. The solar constant is 1350 W m^{-2} at the Earth's orbital radius, the gravitational constant $G = 6.67 \times 10^{-11}$ N m^2 kg^{-2}, the mass of the Sun is 2×10^{30} kg.

16.9   An interplanetary spaceship of 500 kg mass is to be accelerated outwards from a position in free space on the Earth's orbit. It employs a 100% reflecting sail, oriented normal to the Sun's rays, having an area of 1×1 km^2. What outward acceleration can be expected?

16.10   Calculate the radiation pressure on an absorbing target of a stream of 10 g machine-gun bullets fired at 1 km s^{-1} 100 m apart. Solve this problem (a) by the energy density method and (b) by the momentum-transferred-per-second method. Compare the two results and make such comment as seems appropriate.

16.11   An air-raid siren puts out 1 kW of acoustic power in a fairly narrow horizontal beam. Show that the 'recoil' force on the siren and its support is about 3.6 N. For air, $\gamma = 1.4$.

16.12   If the sound beam of Exercise 16.11 has a divergence angle of 15°, show that the radiation pressure 1 km away is 6.7×10^{-5} N m^{-2}.

17

Abstractions as Guides in Physics

Man is all symmetrie,
Full of proportions, one limbe to another,
and all to all the world besides:
Each part may call the farthest, brother:
For head with foot hath private amitie,
And both with moon and tides.

George Herbert
Man

17.1 Introduction

Physicists, especially theorists, try to organize knowledge of physical phenomena over a relatively wide variety of circumstances. They start with a set of assumptions and use accepted principles and rules of procedure devised to analyze and predict the nature or behavior of a specified set of phenomena. When successful, they say they have devised a *theory* which 'explains' the phenomena.

When exploring unfamiliar territory, a physicist welcomes any guidance available to him. Finding analogies or similarities between the behavior you are investigating and behavior you already understand often leads to rapid progress. In fitting a theory to a set of observations or data, usually you need to make some assumptions. Here you may be guided by the caution 'hypotheses must not be multiplied unnecessarily.' This principle is known as Ockham's Razor (after William of Ockham, a fourteenth-century Franciscan friar), and is one of the pointers towards modern science. When we favor a simple explanation over a more complicated one we are expressing the belief that, viewed in the proper way, nature is basically simple. Nature may *not* be simple; the behavior of nuclear constituents, for example, may be forever beyond man's comprehension. However, science can't start from such a defeatist assumption. Furthermore, our considerable success in science is some justification for a more optimistic view of what can be understood.

17.2 Symmetry

Symmetry is one of the dominant abstractions which guide development of theory in physics. We commonly use the term *symmetry* to refer to a 'correspondence in size, shape, and relative position of parts that are on opposite sides of a dividing line or median plane or that are distributed about a center or axis.' Commonly it is the uniformity of *space* which we describe in symmetry properties, but there are extended meanings for the term symmetry which are of great value in physics.

By symmetry we mean that we recognize some property that is unchanged when some other property is changed. For example, the *shape* of a triangle is not changed by *translation* in space, (figure 17.1). A square appears the same when *rotated* through 90° or any multiple of 90°. We say that the square has axial symmetry of shape for increments of 90°. A circle has continuous axial symmetry.

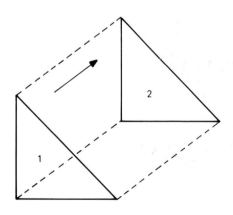

Figure 17.1 The shape of the triangle is not changed by translation from position 1 to position 2.

Consider an equal-arm balance in each of whose pans there is a 10 g brass cube. We should say that this system has symmetry about an axis through the fulcrum. But what if we now change the fulcrum and balance a 10 g cube against a 30 g cube, can we still claim symmetry? Obviously we no longer have symmetry of shape, but we can claim symmetry for a different property: torque (or moment).

In physics we are chiefly interested in symmetries related to conservation laws and symmetries of mathematical functions which lead to physically important predictions. For example, that mechanical laws should have the same form in coordinate systems having uniform relative velocity is a form of symmetry basic to Newtonian mechanics. From symmetry principles we can obtain such practical information as the fact that an isotropic solid has only two elastic constants. In determining the motion of a rigid body subject to arbitrary forces, it turns out to be most convenient to describe the motion in terms of a set of axes fixed in the body. Solution of the equations of motion will be easiest if we follow symmetry considerations in defining the axes. In optics, the properties of a complex lens system can be predicted solely with respect to the symmetries it possesses.

Elsewhere in this book we have examined external similarities, such as an electrical circuit being an analog of heat conduction. We are now looking for internal similarities when we apply symmetry principles to, say, a mechanical system or a nuclear process.

17.3 Symmetry and Conservation Laws

Whenever a description of the laws of physics remains the same when the frame of reference is changed, a symmetry (or invariance) of the physical system exists. For example, changing the origin of the space coordinate system does not change the description of the motion of bodies, for the forces between them depend only on their relative positions. Likewise, a system of bodies behaves the same if it is moved to another place in the original coordinate system. In Newtonian mechanics the law of conservation of momentum is the mathematical consequence of the fact that the basic laws of physics have the same form at all points in space. In the special theory of relativity, however, it is found necessary to require mass to vary with velocity to preserve the conservation of momentum and keep the basic laws symmetrical.

In table 17.1 some symmetries (invariances) are listed along

Table 17.1 Some symmetries (invariances) and conservation laws

Symmetry under:	Quantity conserved
Translation of origin in space (form of basic physical laws same at all points)	Linear momentum, p
Rotation in space (laws of physics have same form in all orientations of axes)	Angular momentum, $\mathbf{L}$
Translation in time (physical laws do not change with time)	Energy, E
Lorentz transformations of special relativity (space–time is isotropic)	Velocity v of the center of mass of a system
Interchange of similar particles	Indistinguishability of similar particles
Certain operations on the quantum mechanical wave function	Electric charge, Q

with the related conservation laws. This subject is still under development. In some cases it is not certain how the conservation law follows from the symmetry principle or how rigorous is the connection.

The first four invariances (symmetries) listed in table 17.1 are termed *continuous* because the changes can be infinitessimal; a finite change can be made bit by bit. The resulting constants relating to the motion are classical quantities and are additive. The other invariances listed refer to quantum mechanics, and there are others. These symmetries (reflections), unlike the classical ones, are discrete: the change considered is not arbitrarily small.

17.4 Symmetry and Parity

Imposing a reflection is equivalent to changing the sign of a coordinate. This may or may not change a mathematical function. The graph of $y = \cos x$ remains unchanged by reflection: reflection means changing $\cos x$ to $\cos(-x)$ and since $\cos(-x) = \cos x$, the appearance of the graph is unchanged (figure 17.2*a*). But the graph of $y = \sin x$ is reversed on reflection since $\sin(-x) = -\sin x$ (figure 17.2*b*). The conserved quantity related to reflection symmetry is called *parity*. The cosine function is assigned *even parity* since it is unchanged by reflection. The sine function is assigned *odd parity* since it is changed by reflection. Changing the signs of *two* coordinates leaves shape unchanged

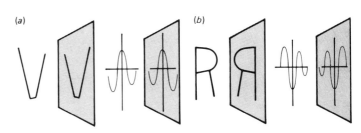

Figure 17.2 Depending on its symmetry, a pattern may be (*a*) unchanged, or (*b*) reversed, by reflection.

and is equivalent to a simple rotation. In three-dimensional space, changing the signs of *three* coordinates causes a reversal. This might be visualized by turning a right-hand glove inside out; by reversing all three coordinates we get a left-hand glove.

We can test the parity of a spinning top by considering its reflection in a plane mirror (figure 17.3). That reflection corresponds to a real experiment: it shows the correct relation among spin **S**, gravitational torque **L**, and precessional **P** vectors. The spinning top therefore has reflection symmetry.

In figure 17.4(*a*) there is an apparent violation of reflection symmetry for the path of an ion moving past a magnetic pole, but are we justified in our use of the symbol N to indicate polarity, in the image? Consider instead that the bar magnet is magnetized by a current (figure 17.4*b*). Reflection in the mirror reverses both this current and the clockwise path of the ion: there is no violation of electrodynamics, and reflection symmetry is satisfied. Consideration of these last two cases, figures 17.4(*a*) and (*b*), should emphasize that we should not be concerned with the reflection of symbols (the N or the arrows) but rather with directly observable quantities which in this case are: the direction of current in the solenoid; the charge and velocity of the moving ion; and the direction of its deflection.

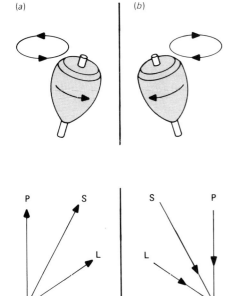

Figure 17.3 (*a*) For the spinning top, vectors show spin **S**, gravitational torque **L**, and precession **P**. (*b*) Its reflection has orientations of **S**, **L**, and **P** consistent with laws of mechanics; the top has reflection symmetry.

17.5 Symmetry and Quantum Mechanics

The term *parity* means space-reflection symmetry or mirror symmetry. The principle of conservation of parity says that one can make no fundamental distinction between left and right. The laws of physics are the same when expressed in a right-hand system of coordinates as they are in a left-hand system. The law holds for all the phenomena described by classical physics, but in 1956 it was shown to be violated for certain types of interaction between elementary particles.

There are three types of interaction between elementary particles: strong nuclear interaction, electromagnetic interaction, and weak nuclear interaction. For the first two, parity is

Figure 17.4 (*a*) There is an apparent violation of reflection symmetry in the path of an ion deflected by the pole of a magnet. The mirror image does not agree with $F = q(V \times B)$. (*b*) Correctly interpreted, reflection reverses *both* the clockwise rotation of the ion *and* the counterclockwise current *I*; there is reflection symmetry.

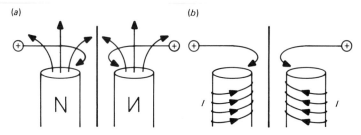

conserved. For example, if a left-polarized particle exists (one whose angular momentum vector is opposite to its linear velocity vector) then there will be an approximately equal number of right-polarized particles. But for weak nuclear reactions, C N Yang and T D Lee suggested that parity is not conserved,[†] and this was later proved experimentally.

The first of two experiments they suggested was a study of the direction of emission of electrons β^- and antineutrinos $\bar{v}$ by a radioactive isotope of cobalt, in the reaction

$$^{60}_{28}\text{Co} \rightarrow \beta^- + \bar{v} + ^{60}_{28}\text{Ni}.$$

In this experiment, performed by C S Wu and her colleagues, $^{60}_{28}\text{Co}$ nuclei were aligned in a magnetic field a few degrees above $0\,\text{K}$ so thermal agitation would not disturb the alignment. Then a count was made of the β particles emitted by the radioactive ^{60}Co in various directions. It was found that there was a strong (probability of electron emission downward (figure 17.5 left); no electrons were emitted directly upward. The antineutrino was most often emitted directly upward, never directly downward. The electron and antineutrino each had a spin of half a unit $(h/2\pi)$. In the decay, the cobalt nucleus had a decrease in angular momentum from 5 to 4 units. So the particles emitted both had to be spinning in the same sense as the nucleus, as shown in figure 17.5(left).

Suppose we always view the cobalt nucleus from the direction in which it appears to spin clockwise. Then in the diagram of the actual experiment (figure 17.5), the electron is emitted towards us. But in the image process the electron would be seen as emitted away from us. This never occurs in the experiment. So the mirror image of a known process is not a possible process. Mirror symmetry is violated in this case (a weak interaction).

The helicity of a particle refers to a screw-like property, specifically the component of the spin vector of a particle along its direction of motion. The neutrino and the antineutrino are

† Lee T D and Yang C N (1956) Questions of parity conservation in weak interactions *Phys. Rev.* **104** (1) 254–8

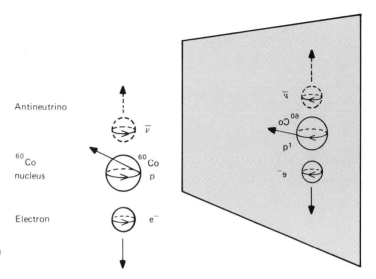

Figure 17.5 Emitted beta particles spin in the same sense as the nucleus.

considered to be mirror images of one another. An analogy is then made: a neutrino is a left-hand screw, the antineutrino is a right-hand screw.

There is reason to believe that if you look at an antineutrino moving away from you it is always rotating clockwise, as would a right-hand screw advancing in this direction. This occurs in the real beta-decay process (figure 17.5), but not for the image process.

17.6 Symmetry and Antimatter

Among several kinds of restricted symmetry not included in table 17.1 is the concept of particle–antiparticle symmetry. For every particle there is believed to be an antiparticle which has the same mass and spin but opposite sign of charge (a *charge-conjugate* particle). For a neutral particle, the antiparticle may be indistinguishable from the particle (such as the photon), or the antiparticle may have opposite helicity to the particle (such as the neutron). Further, for every known process there is another possible process represented by changing each particle into its antiparticle. Ordinary beta decay,

$$n \rightarrow p + e^- + \nu,$$

suggests the process of 'anti-beta decay' in which an antineutron decays into an antiproton

$$\bar{n} \rightarrow \bar{p} + e^+ + \nu.$$

If you are disappointed in the breakdown of reflection symmetry in weak interactions, you may find it satisfying to take the expanded viewpoint of symmetry suggested in figure 17.6. In (*a*) the beta decay of ^{60}Co is represented again. Figure 17.6(*b*) shows every particle changed into its antiparticle. If other properties, including spin direction, remain the same, the process in (*c*) is not possible. The particle emitted upward has right-hand helicity: it cannot be a neutron. In this and many other examples we conclude that all weak interactions violate the principle of particle–antiparticle symmetry. But now note that the process of figure 17.6(*d*), obtainable from (*a*) by mirror reflection plus change of particles into antiparticles, *is* a possible process. The particle emitted upward has left-hand helicity, as a neutrino should. Hence, although each symmetry principle is violated separately, the process is symmetrical for mirror reflection plus particle–antiparticle interchange!

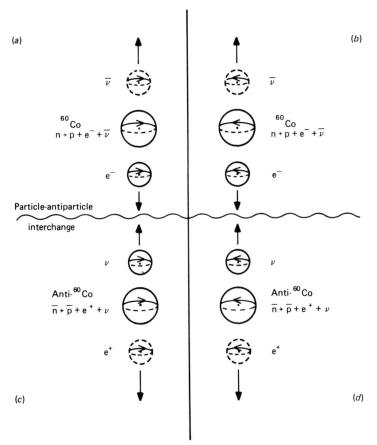

Figure 17.6 (*a*) The beta decay of ^{60}Co; (*b*) reflected in a mirror; (*c*) subject to particle–antiparticle interchange; (*d*) subject to both processes, CP.

Particle–antiparticle interchange is given the short-hand notation C (for Charge conjugation); mirror reflection is denoted P (for Parity). The double operation we have just described is then CP. The process of beta decay of ^{60}Co is said to violate C symmetry and P symmetry but to preserve CP symmetry.

Using quantum mechanics and the special theory of relativity, it seems to be possible to prove that any fundamental process turns out to be symmetrical under the triple operation CPT. By T we mean the operation of time reversal, making each particle retrace its motion as time runs backwards. But now consider some process which violates CP symmetry. The triple operation CPT should result in a possible process. The third operation T has changed an impossible process into a possible one. Following the operation in reverse, time reversal T can turn a possible operation into one that is impossible. So time reversal symmetry appears to have broken down. If so, the laws of nature recognize in a fundamental way the forward direction of time!

17.7 Dimensional Analysis: Models

Another interesting aspect of similarity in physics, and in all sciences, is model making. A model is a representation, maybe pictorial, maybe mathematical, which allows us to think more concretely about some aspect of the physical world. Early in this century Rutherford and Bohr suggested a planetary model of the atom in which electrons travel in orbits about the nucleus, like planets about the Sun. This proved to be an abundantly fruitful model and it still survives, though in a more abstract form.

Closely linked with successful model making are the concepts of similitude and dimensional analysis. In scientific study and engineering design much valuable use is made of similitude: the correspondence in behavior of large and small structures of similar nature. Geometric, kinematic, and dynamic similarities are considered. R C Tolman set forth his view of similitude as a principle: 'The fundamental entities out of which the physical universe is constructed are of such a nature that from them a miniature universe could be constructed similar in every respect to the present universe.'†

P W Bridgman popularized dimensional analysis, earlier suggested by Dupré (1869) and by Lord Rayleigh (1915). Bridgman used similitude in connection with physical quantities having the same dimensional structure, for example R^2 and $4\pi^2 f^2 L^2$ in the expression for electrical impedance.

† Tolman R C 1914 *Phys. Rev.* **3** (2) 244

17.8 Dimensional Analysis

Physics is made quantitative through measurements. The result of a measurement is expressed by a number and a unit or standard of comparison, such as 0.5 kg, 6 ft, or 30 min. Here kilogram, foot, and minute are arbitrary units for the physical properties mass $[M]$, length $[L]$, and time $[T]$. It has been found convenient to define certain arbitrarily chosen properties as *fundamental* dimensions and to treat the others as derived. Thus velocity has the dimension of length divided by time, or $[L][T]^{-1}$. In mechanics, three fundamental properties suffice: the physicist usually takes these as mass, length, and time; the engineer often prefers force, length, and time. When dealing with phenomena not treated in mechanics, it may be expedient to introduce other fundamental quantities. Thus for electricity and magnetism, charge $[Q]$ may be chosen as a fourth fundamental quantity. The derived quantity electrical potential (energy/charge) is then said to have the dimensions $[M][L]^2[T]^{-2}[Q]^{-1}$.

Dimensional analysis provides one check on the correctness of equations. Both sides of an equation must have the same dimensions, as indeed must each term. So $f = \sqrt{ma}$ cannot be true. Much more important, in the absence of a rigorous mathematical derivation, dimensional analysis often enables one to predict the *form* of a desired relation, leaving to experiment the determination of some numerical factor which dimensional analysis cannot provide. Complicated problems in aerodynamics and hydrodynamics have been handled in this way.

17.9 The π Theorem

An expression which remains formally true even when the sizes of fundamental units are changed is said to be a *complete* equation. For a body falling freely from rest, the distance fallen s is equal to $4.9t^2$ only if s is measured in meters and time t in seconds. If s is in feet and time t in minutes the equation is wrong, and it is an incomplete equation. However, $s = \frac{1}{2}gt^2$ is a complete equation for it is valid in any consistent set of units.

Consider a group of n physical quantities, $x_1, x_2, \ldots x_n$ (some of which may be dimensional) for which there is one and only one complete equation connecting them, namely $\phi(x_1, x_2, \ldots x_n) = 0$. Assume that the dimensions of the n quantities are expressed in terms of m *fundamental* quantities, $a, b, c \ldots$. E Buckingham proved that this single relation ϕ can always be expressed in terms of some arbitrary function F of $n - m$ independent dimensionless products, π_1, π_2, ... π_{n-m}, made up from the n variables,

$$\boxed{F(\pi_1, \pi_2, \ldots \pi_{n-m}) = 0.}$$ (17.1)

This is called the π theorem. Even if one does not know ϕ one can often infer the structure of equation (17.1) and thereby gain useful information about the system being examined.

► **Example 17.1** Assume that you do *not* know that $\tau \propto \sqrt{(l/g)}$ and you wish to find the period τ of a simple pendulum.

Choose mass, length, and time as the fundamental quantities, so $m = 3$. Then list all parameters which you think might be relevant to the pendulum's motion, as in table 17.2.
   From table 17.2, $n = 5$, so $n-m = 2$ and we expect to find two independent dimensionless products. Here one is θ, the angular amplitude, another is $l/(\tau^2 g)$. Note that since mass m appears only in the second line of table 17.2, m cannot occur in any dimensionless product. Therefore the period of vibration τ does not depend on the mass of the bob and the π theorem gives $F(\theta, l/\tau^2 g) = 0$, or

$$\tau = \Theta(\theta)\sqrt{(l/g)},$$ (17.2)

where $\Theta(\theta)$ is an arbitrary function of θ. If we make an additional assumption that θ is small enough to be neglected, then $n = 4, n - m = 1$ and $F(l/\tau^2 g) = 0$. For this to be true for all values of the variables

$$\tau \propto \sqrt{(l/g)}.$$ (17.3)

The π theorem thus predicts that the period of the pendulum is proportional to the square root of its length and inversely proportional to the square root of the acceleration due to gravity. The constant of proportionality (actually 2π) cannot be obtained from dimensional analysis.

   To link our simple example, above, to model making, suppose that before constructing a huge, expensive pendulum for display at a World Fair you wanted to determine its period from a model of, say, $1/100$ its length. Since g would be the same for both the

Table 17.2   Parameters relating to a simple pendulum

Quantity	Symbol	Dimensions
Length of string	l	$[L]$
Mass of bob	m	$[M]$
Acceleration of gravity	g	$[LT^{-2}]$
Angular amplitude	θ	$[0]$
Period of swing	τ	$[T]$

large pendulum and its model, equation (17.3) predicts that the period of the model would be just $1/10$ that of the large pendulum.

17.10 Dimensionless Groups

Any combination of quantities which possesses zero overall dimensions is called a dimensionless group. Such groups are important in engineering studies of complicated processes and as criteria for similarity in model studies. Scores of such dimensionless groups have been tabulated; many are named after pioneers in the field. Important in fluid mechanics, for example, is the Reynolds number $N_{Re} = vD\rho/\eta$ where v is the velocity of a body through a fluid having viscosity η and density ρ and D is some length (diameter) defining the size of the body.

Among the advantages of using dimensionless groups in studying complicated systems are (a) reduction in the number of 'variables' to be determined, (b) simplification of the scaling-down or scaling-up of results obtained with models, and (c) attainment of results independent of both the scale of the model and the system of units used.

17.11 Models in Physics

Models of physical phenomena often make some simplifying assumptions, then set up a picture or mechanical model as an aid in understanding the phenomenon and its dependence upon certain relevant physical quantities. A model may be pictorial, such as Atlas carrying the world on his shoulders, or the kinetic-theory picture of a gas as being made up of tiny elastic molecules in rapid random motion. A model may be representational, for the purpose of demonstration or experiment, as was an early planetarium to interpret the significance of planetary motion. Sometimes a model may be consciously provisional in nature: an oversimplification to picture a complicated physical phenomenon for which an adequate representation is not known. Examples of this include models of chemical binding and models of electrical conduction in metals.

17.12 Models in Engineering

Our theme of similarities is underscored in construction of small-scale models used to obtain necessary engineering information for the construction of a large, complicated, and expensive structure. The structure may be an aircraft, a flood-control system, a concert hall, or a telephone network. Consideration of dimensional analysis shows the effect of scaling down some variables, say, size and mass, while others such as viscosity remain constant. Dimensional analysis may show that it is impossible to get exact information about the behavior of the proposed structure from a scale model, but it may also suggest ways of balancing the effects of two variables, or ways of mathematical approximation to obtain useful information.

17.13 Operators and Operations

Science is often and rightly regarded as a discipline for the systematizing of knowledge and the uses we may make of it. It is therefore natural that we should seek some way of drawing together all problem-solving activities and placing them within a common formal construct. All your life you have been solving problems of one sort or another, from the numerical and symbolic situations of mathematics and science, to the design and construction of something you need or want. But have you ever thought that problem-solving, varied and ramified though it is, can be encompassed, conceptually at least, by a single philosophical wraparound?

The solving of a problem can be thought of as the finding of a pathway from a particular element of the universe of all possible problems P_m to a one or more particular elements of the universe of all possible solutions S_n. The pathway lies through what we shall call an operations matrix, containing all possible things you might do in attempting to solve the problem, from multiplying by 48 to postponing your vacation. This generalizing concept is illustrated in figure 17.7.

Of all the possible solutions, the only ones we're interested in are those which satisfy the initial conditions of our problem. These we shall call valid solutions. The set-piece problems you find in textbooks are usually designed to have just one valid solution, however sometimes they have more than one valid solution. On the other hand, for some problems, completely valid solutions may not exist at all and a choice must be made among the least undesirable of several approximate solutions.

Usually there will be alternative pathways through the operations matrix from a problem input to a valid solution. One of these alternatives is indicated by the broken path in figure 17.7. In most cases, however, one of the pathways will be clearly preferred

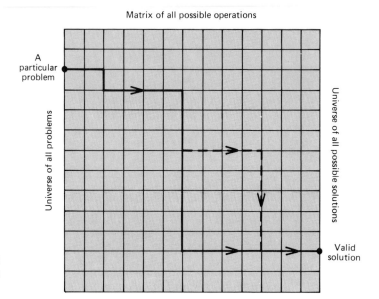

Matrix of all possible operations

A particular problem

Universe of all problems

Universe of all possible solutions

Valid solution

Figure 17.7 There may be more than one path from a particular problem to a valid solution.

because it involves a minimum number of operations or because it satisfies some self-imposed criterion of 'elegance.' Thus the answer to the problem 'Find the product of sixteen and four' can be obtained in a single step by applying to the number sixteen the operation of multiplication by four. The same valid solution could be obtained alternatively by the pathway of multiplying sixteen by five and then, in a separate operation, subtracting sixteen. The people who design flow-charts for computer programs are very familiar with consideration of alternate pathways.

The simplest problems are those having valid solutions obtainable by applying a single operation to the input data (e.g. sixteen times four). The more challenging problems, of course, are those requiring the application of several operations, often in a unique sequence prescribed by the rules of logic, as in the proof of a theorem in geometry.

The individual elements of the operations matrix are the operations themselves. An operation is anything you do to something to obtain a result. An operation must have an input—the thing that is operated upon, sometimes called the operand. It also has an output, which is the result of an operation on the operand by the operator. In table 17.3 are some of the operators and operations you have frequently encountered.

Note that some of the operators listed are material objects, devices, or systems, such as amplifiers, factory machines, transformers, etc. Other operators are mathematical and are brought to bear by symbol manipulation on paper. A mathematical

Table 17.3 Some operators and operations

Operand	Operator	Operation	Output
Raw materials	Factory processing	Manufacturing	Finished goods
Water	Heat	Boiling	Steam
$\begin{cases} \text{Any number,} \\ n \end{cases}$	Multiplier n	Squaring $n \times n$	Squared number n^2
Input power at I_1, V_1	Transformer	Transformation	Output power at I_2, V_2
$\begin{cases} \text{Electrical signal} \\ V = f(t) \end{cases}$	Linear amplifier $\times 100$	Amplification (multiplication)	Enhanced replica $V = 100 f(t)$
$\begin{cases} \text{Sine wave} \\ A e^{i\omega t} \end{cases}$	Phase shifter $e^{i\theta}$	Phase shifting (multiplication)	Phase shifted wave $A e^{i(\omega t + \theta)}$
$f(x)$	$\dfrac{d}{dx}$	Differentiation	Slope of function $f(x)$
$f(x)$	$\int_{x_1}^{x_2}$	Integration	Area under curve
$\begin{cases} \text{Principal sum} \\ \$_0 \end{cases}$	Time at specified interest rate r $\$_0 (1 + r^n)$	Appreciation (multiplication)	Present value $\$$
House	Paint, brush, labor	Painting	Painted house
$\begin{cases} \text{Sine wave} \\ V = A \sin 2\pi f t \end{cases}$	Frequency doubler	Frequency doubling	Sine wave of twice the frequency $A \sin 4\pi f t$

operator is any command which tells a person or a calculator how to process the data.

The function of a linear amplifier sitting on the workbench is to perform the operation of strengthening the amplitude of a real input voltage signal to produce an output replica of the signal having identical wave shape and larger amplitude. The symbolic representation of this process would be to take an input signal $V = f(t)$, apply to it an operation called multiplication by a scalar number, N, to obtain an output signal $V = N f(t)$. The real operation and its symbolic analog are indicated by the bracketed pairs in table 17.3. Although only a few such pairs are shown, any operator and operation can be represented by suitable symbols.

The great utility of mathematical operations in application to the sciences is that real-world operators and operations can be simulated on paper by analogous mathematical operators and operations to obtain solutions to problems without continually having to go into the field or into the laboratory and actually perform the experiments. How costly and wasteful it would be for an engineer when he asks 'How strong must I design the cable of this suspension bridge in order for it to carry the expected load?' if he actually had to go and *build* a test bridge to *see* whether it would carry the load. How could we have possibly got to the

Moon if we couldn't have calculated beforehand the exact program for the journey? Remember, nobody had been there before. We *had* to know what was going to happen before the process was undertaken. We had to have confidence that mathematics would represent the real situation.

In examining points of contact between sophisticated pure-mathematical concepts and potential scientific applications, Browder† remarks 'To use an industrial metaphor, mathematics (and especially pure mathematics) is the machine-tool industry of the sciences, and we cannot use only the tools manufactured in the past unless we believe that science will never again have significantly new problems to solve—a belief that is clearly false. . . . The potential usefulness of a mathematical concept or technique in the advance of scientific understanding has very little to do with what we can foresee before the concept or technique has appeared.'

To illustrate this point, Browder reminds us that in 200 BC the Greek geometer Apollonius of Perga wrote his celebrated *Treatise on Conic Sections*—an exercise in pure mathematics since no applications of his results were considered or made in the classical world. But in 1604, Kepler read the writings of Apollonius and applied them in optics and the study of parabolic mirrors. In 1609 Kepler made the brilliant observation that the orbits of the planets should be described as ellipses, not by means of circles and epicycles, thereby laying the principal foundation for Newton's later theory of gravitation. Currently, applications of mathematical concepts are being made much more rapidly, for example in matrix mechanics, wave mechanics, and automata.

Problem-solving, then, consists of applying appropriate operations, real or symbolic, to the problem input in order to obtain an output fulfilling the specifications of a valid solution. The next question is one to try men's souls: 'How do we select the particular operation or sequence of operations that will result in a valid solution?'

Most obvious, of course, is the random-trial approach, in which one simply applies operation after operation, singly or in various random sequences, in the hope of finding an output which fits the specifications of the problem. In view of the enormous number of possible operations and operation sequences, however, it must be evident that the probability of finding a valid solution by chance is vanishingly small. There must be a better way.

† Browder F E 1976 Does pure mathematics have a relation to the sciences? *Am. Sci.* **64** 542–9

Sometimes the statement of a problem indicates the operation or operations required in its solution. The problem 'What is sixteen times four?' clearly tells you that multiplication is the appropriate operation. But what about the more complicated problems for which helpful clues are absent? Here's where the ingenuity and perspicacity of the problem-solver must be brought to bear. The task is often simplified by the recognition of similarities, the subject of this book. Problems seldom occur in isolation but rather come in families. When one problem of a family is solved, we may logically expect that the application of a similar sequence of operations to other members of that family will result in valid solutions.

Sometimes a problem which is not recognizable as a member of a family can be recast, with a little preliminary manipulation, so as to endow it with a family association. Dr Frank Kocher has suggested the following: consider the situation of a room empty except for a hot stove, a table, and a kettle of cold water resting on the floor. The problem is how to heat the water, and the solution is to pick up the kettle from the floor and place it on the stove. Suppose we then leave the room and return later to find the kettle, with water now cold, resting on the table instead of on the floor. Almost everybody will say, 'Pick up the kettle from the table and set it on the stove,' but the occasional sharpie alert for similarities will say, 'Take the kettle from the table and place it on the floor, thus reproducing the initial conditions of a problem we have already solved. From there, proceed as before.' Similarly we recall how Descartes devised a solution for the general quartic equation. He introduced a substitution of the variable which resulted in a reduced quartic for which a standard solution was already known.

▶ **Exercises**

17.1   A quantity is *invariant* if it has the same value in different inertial systems of reference. A quantity is *conserved* if it remains unchanged in time. Can a quantity be invariant and yet not conserved? Show that energy is conserved and is invariant. Show that the rest (or invariant) mass m_0 of a system defined by $E^2 - (pc)^2 = (m_0 c^2)^2$ is both invariant and conserved. Here E is the total energy in the laboratory frame, p is the magnitude of total momentum in that frame.

17.2   The decay of a certain particle A into two other particles, A → B + C, would violate no known laws of physics except that the rest mass m_0 of A is smaller than the sum of the rest masses of B and C by the amount

Δm. Is this process possible if A moves through otherwise empty space with kinetic energy greater than $(\Delta m_0)c^2$?

17.3   Can you suggest a way of determining, from Earth, whether a distant star is made of matter or antimatter? Assume that you have available for examination light from the surface of the star and also neutrinos emitted in the nuclear reactions which provide the stellar energy, whose net effect is

$$_1^1\text{H} + _1^1\text{H} + _1^1\text{H} + _1^1\text{H} \rightarrow _2^4\text{He} + e^+ + e^+ + \nu + \nu.$$

17.4   Temperature is always measured by indirect experimental observations; there is no physical process for addition of temperatures. A temperature near $0\,\text{K}$ is determined by a magnetic experiment. The temperature on the 'magnetic' scale may not be on the same scale as a plasma temperature determined by an optical pyrometer. How then can you justify using *temperature* as a *fundamental* quantity in dimensional analysis?

17.5   Physicists and engineers have found dimensional analysis a welcome aid especially in problems in mechanics, hydrodynamics, aerodynamics, and heat transfer. Why then, do you suppose, when it comes to electrical and magnetic magnitudes, have discussions of dimensions been characterized as 'richly endowed with meaningless verbosity,' by Focken as 'a big, buzzing, blooming confusion,' and by Bouasse as a 'collection of stupidities?' (These quotations do not represent the final word on the value of dimensional analysis.)

17.6   For a one-dimensional temperature field $\partial T/\partial t = a\partial^2 T/\partial x^2$, where T denotes temperature, and t time. The distribution of potential in a conductor is described by

$$\frac{\partial V}{\partial t} = \frac{1}{RC}\frac{\partial^2 V}{\partial x^2}.$$

The so-called Fourier number at_0/l^2 is the analog of t_0/RCl^2. In surveying temperature variations in soil, near the surface and to a depth of about $10\,\text{m}$, why would you recommend use of an electrical analog? Suggest a method.

17.7   Add three more operators and operations to the partial list of table 17.3.

17.8   How do you react to the following assertion? 'There is nothing in the solution of a problem which is not already implicit in the input information.'

17.9   The assignment of dimensions is purely conventional. (a) Give the dimensions of moment of inertia of a mass in the mass–length–time system, and (b) in the force–length–time system. (c) Consider a system S in which the unit of work is the energy equivalent of a 1 g mass, from $E = mc^2$. Take mass and time as fundamental dimensions, with units of gram and second. If Newton's equation $F = ma$ is retained, what are the dimensions of length, velocity, and force in system S? (d) Prove that the unit of length in system S is the distance traveled by light in $1\,\text{s}$.

17.10   When the plates of a charged capacitor are suddenly connected by a conductor, the period T with which charge oscillates between the plates depends on the resistance R and the self-inductance L of the conductor and on the capacitance C of the capacitor. (a) Use dimensional analysis to find the general form of the equation for T. If L is negligible, (b) how does T vary with C and (c) how does it vary with R?

17.11   From dimensional analysis, find the acceleration a of a point moving in a circle at a constant speed.

17.12   Use dimensional analysis to find the speed of propagation of a wave in deep water. Since gravitational force is predominant you can ignore surface tension and viscosity. Assume that the depth is too great to influence the speed.

17.13   Lord Rayleigh proposed treating the vibration of a star, a fluid body held together by its own gravity, using dimensional analysis. Assume that the variables that can affect the frequency f of a natural mode of vibration are diameter D, density ρ, and gravitational constant k. Show that the frequency is independent of diameter and directly proportional to the square root of density.

17.14   The water in a river 20 m wide is flowing with a speed of $2\,\mathrm{m\,s^{-1}}$ around a bend of 100 m radius. How much higher is the water level on the bank at the outside of the curve than it is on the inside bank? Identify all the input data. List in logical sequence, and describe, the separate operators and operations required in the solution of this problem.

18 Here and There with Similarities

An isolated fact can be observed by all eyes; by those of the ordinary person as well as of the wise. But it is the true physicist alone who may see the bond that unites several facts among which the relationship is important though obscure.

Jules Henri Poincaré

18.1 Introduction

In previous chapters we have dealt with numerous cases where similarities exist among phenomena in different areas of physics and engineering. The cases we selected were chosen to be illustrative rather than exhaustive, and many other opportunities to point out similarities were, of necessity, passed over. We wish, in this final chapter, to mention a few of what seem to us to be the more interesting of the similarities not yet discussed.

18.2 Thoughts on Thermodynamics

Thermodynamics is the study of the relationships between heat and other forms of energy, notably work. Work is done in many physical systems. The amount of work done, or the amount of energy involved in a process, is often calculated as the product of two quantities. In linear mechanics, for example, these two quantities are force F and distance Δx. In rotational mechanics they are torque τ and angular displacement $\Delta\theta$. The work obtainable from a weight mg dropping through a gravitational potential difference Δh is the product $mg\Delta h$. Similarly, to obtain energy from an electric charge q we cause it to fall through an electrical potential difference ΔV, obtaining energy in the amount of $q\Delta V$. To obtain work from a compressed gas at pressure p we allow it to expand through a volume potential difference Δv, and the work done is $p\Delta v$. Gathering these examples and adding a few more, we can construct a table of similarities (table 18.1).

Note that in the first four physical situations of table 18.1 the work done *by* the system must previously have been performed *on* the system, residing meanwhile in the system as potential energy.

Table 18.1

Physical system	First quantity	Second quantity	Work done or energy involved
Falling weight	mg	Δh	$mg\Delta h$
Unwinding coiled spring	τ	$\Delta\theta$	$\tau\Delta\theta$
Expanding gas	p	Δv	$p\Delta v$
Electric circuit	q	ΔV	$q\Delta V$
Object pushed across floor	cmg†	Δx	$cmg\Delta x$
Ship moving through water	F‡	Δx	$F\Delta x$

† c is the coefficient of friction
‡ F is the propeller thrust

In the last two situations work is being done *on* the system, but there is no mechanism for storing it for subsequent release. Work done against friction is wasted as heat, and work done while pushing a ship through the water is wasted in creating turbulence, overcoming viscous friction, and producing waves.

With the similarities of table 18.1 in mind, it would be easy to extend the table by saying, naïvely, that to obtain work from heat we should take a quantity of heat H and allow it to fall through a thermal potential difference ΔT, obtaining work in the amount $H\Delta T$. However, when you go through the thermodynamic reasoning leading to the Carnot efficiency, you find that the greatest amount of work you can obtain from a given amount of heat H taken from a heat source at Kelvin temperature T is

$$W = \frac{H\Delta T}{T}. \tag{18.1}$$

You enquire, 'Why is the orderly panorama of energy expressions in the last column of table 18.1 violated in this particular case? What's so exceptional about heat that it has to have a special formula *not* similar to the others?'

The thing that is different about heat is that it does not have the dimensions suitable for inclusion in either the second or third column of table 18.1. Heat already has the dimensions of energy.

The heat example is not really anomalous. Consider the analogous situation in mechanics of a mass m resting on a mountain top waiting to roll down into a neighboring valley, doing work as it goes. On the mountain top, distant r_1 from the center of the Earth, the mass has potential energy $\int_0^{r_1} mg(r)\,dr$. This energy may be considered as similar to a quantity of heat H_1 at source temperature T_1. After the mass has rolled into the valley at distance r_2 from the center of the Earth, its potential energy is $\int_0^{r_2} mg(r)\,dr$. The amount of work it could have done is the difference between the two integrals. Even though the mass has gone as far as it can go, it still has potential energy $\int_0^{r_2} mg(r)\,dr$ which is unavailable for further conversion into work because the valley bottom is as close to the center of the Earth as the mass can get. This residual unavailable energy is like the heat H_2 rejected by the heat engine into the heat sink, because the temperature T_2 of the sink is as close to absolute zero as the engine's condenser can attain. These similarities are sketched in figure 18.1.

You may have heard or read the assertion, 'You can convert work into heat with 100% conversion efficiency, such that for every 4.19 J of work converted, 1 calorie of heat is produced. But you can *not* convert 1 calorie of heat back again into 4.19 J of

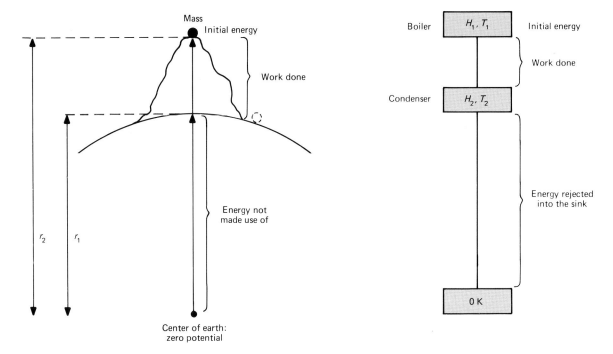

Figure 18.1 Obtaining work from a descending mass is akin to getting work from a heat source.

work with any physically realizable heat engine.' This statement is rather loose; let's try to tighten it up. It is true that to obtain work from heat you must 'degrade' it through a temperature drop. Imagine a heat engine working between a heat-source temperature of 600 K and a condenser temperature of 300 K. This engine can in principle abstract two calories of heat from the source, do *one* calorie's worth of work, and reject one calorie of heat into the sink. In this context the quotation above is correct; but the 1 J of heat that 'disappears' in the process *is* converted 100 % into work.

Figure 18.1 suggests an analogy between the 'heat death' of the Universe and what we might call the 'gravity death' of the Earth. When all sources of heat in the Universe have been exhausted, there will be no temperature differences from which to derive mechanical work by means of heat engines. The Universe will then have the same energy content it now possesses, but the energy will be unavailable to do any work. Similarly, if all the mass now occupying mountain tops on the Earth were eventually transferred to fill up the valleys, there would be no more mountainsides for masses to roll down. The Earth would then become dead as far as obtaining work from falling masses is concerned. It would still have an enormous amount of potential energy, but this energy would be unavailable.

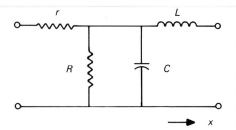

Figure 18.2 An AC transmission line is a succession of these unit sections.

18.3 Propagation of Waves in Lossy Media

(i) AC Transmission Line In §4.4, we considered the direct current in a long two-wire transmission line having both series resistance along and shunt resistance between the wires. When we consider the transmission of alternating current, we find that other properties of the transmission line must also be taken into account. Figure 18.2 shows that for AC propagation a typical transmission line consists of a sequence of unit sections having series resistance r, shunt leakage resistance R, series inductance L, and shunt capacitance C per unit length. While the figure shows these properties as represented by discrete circuit elements, they are in actuality distributed uniformly along the line.

If an alternating current $i = i_0 \sin 2\pi f t$ is introduced into such a line at $x = 0$, it can be shown† that the current at any point x along the line is given by

$$i = i_0 e^{-\Gamma x}, \tag{18.2}$$

where Γ is the propagation constant. In turn, Γ is composed of two other quantities α and ϕ, where α is the attenuation constant and ϕ is the phase constant per unit length: $\Gamma = \alpha + i\phi$. It can further be shown that

$$\alpha = \frac{1}{\sqrt{2}} \sqrt{\left(\frac{r}{R} - 4\pi^2 f^2 LC + \sqrt{[r^2 + 4\pi^2 f^2 L^2)(R^{-2} } \right.}$$

$$\left. + 4\pi^2 f^2 C^2)] \right),$$

$$\phi = \frac{1}{\sqrt{2}} \sqrt{\left(\frac{-r}{R} + 4\pi^2 f^2 LC + \sqrt{[(r^2 + 4\pi^2 f^2 L^2) \times } \right.}$$

$$\left. (R^{-2} + 4\pi^2 f^2 c^2)] \right). \tag{18.3}$$

In the construction of practical transmission lines, R can be made very large and L can be made small enough so that the inductive reactance per unit length is negligible compared with the capacitive reactance per unit length at the frequency of operation. The propagation characteristic of the line is then dominated by r and C. This is particularly true of submarine cables where the conductors can not be separated by large distances to reduce the capacitance. Setting R equal to ∞ and L

† Page and Adams *Principles of Electricity* (New York: Van Nostrand), Chap 15, or Smythe *Static and Dynamic Electricity* (London: McGraw-Hill) Chap 10

equal to 0 in equation (18.3), we obtain

$$\alpha = \phi = \sqrt{(\pi f r C)},$$

so that

$$i = i_0 e^{\sqrt{(\pi f r C)}} \sin\left[2\pi f t - x\sqrt{(\pi f r C)}\right]. \qquad (18.4)$$

This expression shows that the magnitude of the current decreases exponentially along the line and suffers a phase lag given by $x\sqrt{(\pi f r C)}$. The physical meaning of this result is that the current is no longer the same in all parts of the circuit, but is progressively 'used up' in charging and discharging the capacitances of the successive unit sections of the line.

The exponential decay indicated in equation (18.4) is more severe for higher frequencies, depending on f. This fact was of significance in early submarine cable transmission. The dots and dashes fed into the cable had steep rises and sharp corners. However, the higher-frequency Fourier components of the dots and dashes traveled at different speeds and attenuated more rapidly with distance, so that the square corners were missing from the received signals; a dot or dash was reduced to a barely recognizable smear. The modern submarine cables now used for voice transmission have been specially engineered with series inductance added to make them distortionless over the band of frequencies used for voice channels.

(ii) Thermal Waves in the Earth Let us now switch to the consideration of an analogous problem in heat transmission. We wish to analyze the penetration into the Earth of daily and yearly temperature fluctuations. The surface of the ground is heated during the day and cooled during the night with a temperature cycle that can be considered to be essentially sinusoidal: $T = T_m + T_0 \sin 2\pi f t$, where T_m is the mean temperature and T_0 is the peak amplitude of the fluctuation at the surface. These temperature fluctuations penetrate some distance into the Earth.

The heat-flow equation which must be satisfied in the solution of this problem is

$$\frac{\partial T}{\partial t} = \frac{\kappa}{\rho c}\frac{d^2 T}{dx^2}. \qquad (18.5)$$

This equation is a statement of continuity. It says that the rate of change of the temperature of a volume element is proportional to the rate at which heat is flowing into it, with the constant of proportionality equal to $\kappa/\rho c$, where κ is the thermal conductivity of the Earth material, ρ its density, and c its specific heat capacity. The quantity $\kappa/\rho c$ is called the thermal diffusivity of the medium. Often we need to describe the propagation of temperature T in a body whose surface is periodically heated and cooled. That is, we need a solution of equation (18.5) for the boundary condition

$$T_{x=0} = T_m + T_e \sin(2ft).$$

According to a solution of equation (18.5) developed by Lord Kelvin and used by him in his studies of the propagation of temperature waves in the Earth, the temperature at any depth x is given by

$$T = T_m + T_e \exp\left[-x\sqrt{(\pi f \rho c/\kappa)}\right] \sin\left[2\pi f t - x\sqrt{(\pi f \rho c/\kappa)}\right].$$

$$(18.6)$$

This equation shows that the temperature fluctuation amplitude decreases exponentially as we go deeper into the Earth and suffers a phase lag of $\sqrt{(\pi f \rho c/\kappa)}$ per unit length. The similarity between equation (18.6) and equation (18.4) justifies our considering the Earth as a transmission line for temperature waves. In particular, we can see that the thermal conductivity κ of the Earth corresponds to the reciprocal of the series resistance r of the electrical line, and that the thermal quantity ρc, which signifies heat capacity per unit volume, corresponds to the capacitance C per unit length of the electrical line. We remark here that heat conduction involves no property corresponding to electrical inductance, and that the transmission lines we considered in this section were taken to have negligible inductance. The analogy between the thermal and electrical models is therefore a valid one.

Equation (18.6) shows that the exponential decay of the amplitude of the thermal fluctuation with depth depends upon the frequency of the fluctuation at the surface; the higher the frequency the more the attenuation. This prediction is borne out in nature. Annual fluctuations can be detected down to about 20 m in the Earth, while daily fluctuations are barely detectable deeper than 1 m.

(iii) p–n Junction   In another sudden switch we turn to p–n junction theory. In the operation of a p–n junction diode the current is carried by electronic particles which become minority particles† after they have crossed over the junction barrier. These minority particles diffuse away from the junction, recombining with normal majority particles as they go, so that the minority particle concentration decreases with increasing distance from the junction, thus causing the concentration gradient which

† A minority particle is an electron in a p-type semiconductor or a positive hole in an n-type semiconductor.

drives the diffusion current. If an alternating current $i_0 \sin(2\pi ft)$ is applied across the junction, the concentration of minority particles just across the barrier will be approximately $c_0 \sin(2\pi ft)$. The concentration c of minority particles at a distance x further from the barrier will be

$$c = c_0 e^{-gx} \sin(2\pi ft - hx), \qquad (18.7)$$

where

$$g^2 = \frac{1}{2}\left[\frac{1}{D\tau} + \sqrt{\left(\frac{1}{D^2\tau^2} + \frac{4\pi^2 f^2}{D^2}\right)}\right],$$

and

$$h^2 = \frac{1}{2}\left[-\frac{1}{D\tau} + \sqrt{\left(\frac{1}{D^2\tau^2} + \frac{4\pi^2 f^2}{D^2}\right)}\right].$$

In these equations D is the diffusion constant for the minority particles and τ is their average lifetime before disappearing by recombination. The similarity between equations (18.7), (18.6), and (18.4) is at once apparent. So we can regard a semiconductor in the neighborhood of a p–n junction as a lossy transmission line for minority carriers that have crossed the barrier.

Diffusion is a process which, like heat transfer, has no property corresponding to inertia in mechanics or inductance in electricity. Equations (18.7), (18.6), and (18.4) are solutions to similar second-order partial differential equations. These are sometimes referred to as diffusion-type equations to distinguish them from 'ordinary' wave equations, of which equation (7.1) is an example and in which inertia or inductance is involved.

Is it appealing to your fancy to consider that the current in a 3000 mile long submarine cable should exhibit the same physical behavior as the concentration of minority particles in a 10^{-3} cm thick layer of a semiconductor? Such is the nature of nature.

18.4 Feedback Control

Scottish engineer and philosopher James Watt improved the steam engine in numerous ways. In about 1800 he invented the centrifugal governor by which the speed of a rotating engine is automatically controled. In this device (figure 18.3) a pair of masses C are attached to arms hinged on a vertical shaft S which rotates at a speed proportional to that of the engine. If the rotational speed increases, the increased centripetal force on the rotating masses is provided by outward motion of the hinged

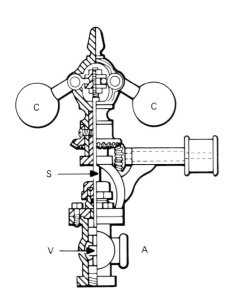

Figure 18.3 The speed of a steam engine is held constant as the centrifugal governor adjusts the steam valve V.

arms. This movement depresses spindle S and moves valve V to decrease the supply of steam. When the speed of the engine decreases, the balls descend, opening the throttle. So the speed of the engine is kept reasonably constant under varying loads.

The governor is an example of *feedback*, the return of 'output' signals to the 'input' of any device for the purpose of correcting or improving the operation of the device. The governor is a case of *negative* feedback for it *decreases* the steam supply with greater shaft speeds. If the centrifugal device had been installed incorrectly so as to supply more steam as the speed increased, we would have positive feedback resulting in a rapid speed-up which might destroy the machine.

An even older example of feedback is a question-and-answer conversation. If only one person does all the speaking, there is no assurance that the second person has received and understood the intended message, but if two persons continually monitor one another, a critical and scientific conversation becomes possible.

Mathematician Norbert Wiener in 1948 coined the word *cybernetics* to designate the theory of 'control and communication in the machine and in the animal.' The idea of feedback is central to this theory as it also is in *automation*. This latter term was invented by the American industrialist D S Harder in 1936, to mean the application of mechanical or electronic techniques to minimize the use of manpower in any process.

(i) Closed-loop Control   A distinction is made between open-loop and closed-loop operations. An *open-loop* system is one in which the control action is independent of the output or the desired result. Suppose you want to schedule the temperature of a glass-annealing furnace so that the temperature will rise to a given value, remain constant for some minutes, then decrease, slowly at first and then more rapidly. You could design a cam which when rotated slowly by clockwork would press against a spring-loaded gas valve to admit the proper amount of fuel to the furnace at the proper time to achieve the desired temperature–time schedule, but if someone opens the windows, if gas pressure changes, or there is some other disturbance, your open-loop control system can not respond and will not achieve the desired results.

A *closed-loop* system is one in which the control action is dependent upon the output. A simple example is the temperature control in a home heating system (figure 18.4). At the thermostat the control gap between the contacts is set by adjusting the screw according to the temperature desired. The bimetallic element measures temperature through the differential thermal expansion

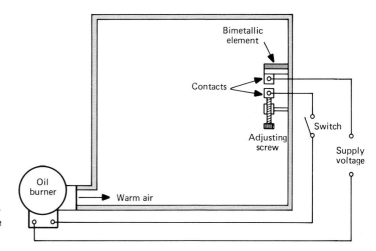

Figure 18.4 Room temperature is controled by the simple on–off action of the thermostat.

of two dissimilar metals. The element shown is designed to bend down when the temperature drops. When room temperature cools below the set value, the contacts close and if the switch used to actuate the heating system is also closed, the oil burner will start up and send warm air into the room. When the room heats above the set temperature, the bimetal bends upward, opens the contact, and shuts off the burner. Should the room again cool below the set temperature the contacts close and return the temperature to the desired value. Thus the room temperature is maintained close to the desired value regardless of changes in outdoor temperature, wind, or other conditions.

A closed-loop or feedback control system has a sensor which measures the quantity to be controled, and a controler which compares the actual value with the desired value and if they differ institutes corrective action (figure 18.5). A process in which a

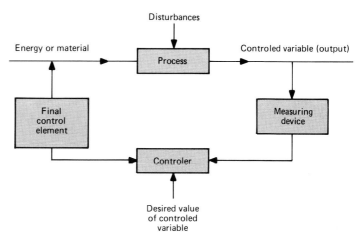

Figure 18.5 When the desired value of the controled variable is set in the controler, the effect of any disturbance is neutralized by the feedback control loop.

person provides the control also may be regarded as a closed-loop system when that person determines the error (set value minus actual value) from an indicator and then makes the appropriate correction.

(ii) Some Types of Control   The examples given so far are of simple on–off control, but suppose you were given the task of holding the water in a tank at 50 °C by adjusting the flow of steam into it (figure 18.6). With the steam shut off, the water temperature would be about 10 °C; with the valve fully open the water temperature would be about 80 °C. You would probably note any discrepancy between the indicated temperature and the desired temperature, 50 °C, and then if necessary change the valve setting, a small amount for a small discrepancy, a larger amount for a larger discrepancy—not a simple on–off control. With experience, you might also let the *rate* at which the temperature was changing influence your setting of the valve. These and other modes of control are often incorporated into automatic control systems, responding more rapidly and precisely than a human operator could.

The thermostat of figure 18.4 has a two-position action (on–off), ideally at a single temperature. In another type of two-position action, there is a *differential gap*: the output (controled) variable is allowed to wander within a certain range on each side of the set value; corrective action is initiated only when the output moves beyond this gap.

In *proportional-position* action there is a continuous linear relation between the value of the controled variable and the position of the control element (such as the valve in figure 18.6). In *floating-speed* action the control element is moved at a fixed rate between its extreme positions. An obvious refinement is *multi-speed floating* action in which the control element is moved

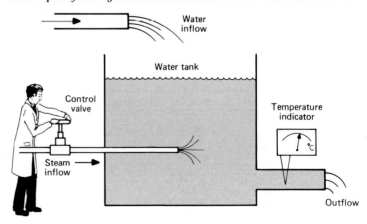

Figure 18.6 Water in the tank is held near a constant temperature by adjusting the flow of steam.

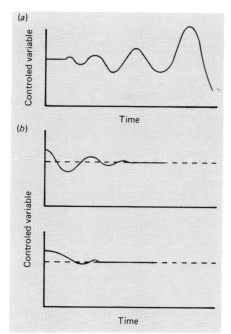

Figure 18.7 The system represented in (*a*) is unstable, and it oscillates with increasing amplitude. The systems represented in (*b*) are relatively stable, and oscillations in the controled variable quickly die down.

at two or more rates, each corresponding to a definite range of values of the controled variable—faster correction being applied for larger errors. When proportional-speed floating action is combined with proportional-position action, the former is said to become *reset* action. There are even more sophisticated control actions such as *rate control* which is a second-derivative control action. Generally one adopts the simplest control action which satisfies the needs of the process to be controled.

(iii) Stability and Accuracy

We may loosely define the *sensitivity* of a control system as the amount of corrective effort applied per unit error. *Instability* is the tendency of the controled variable to oscillate either between definite bounds or with unrestrained amplitude, even when the error has become zero. Too high a sensitivity leads to instability. Instability is related to time lags, in the measuring element, in the controler, and in the process, (for example the time for a water bath to absorb the heat supplied).

Figure 18.7(*a*) shows the behavior of an absolutely unstable system. Systems which respond as shown in figure 18.7(*b*) are said to be relatively stable: on being disturbed the systems go to a new equilibrium with damped oscillations. The system described by the lower curve has greater stability than the other.

We say that we increase the accuracy of a feedback control system when we decrease the errors, both transient and steady state. Increased accuracy may often be achieved in increasing the sensitivity of the system, but a compromise must often be made since increasing the sensitivity tends to render the system less stable.

(iv) Steering Servomechanism

Feedback control was first applied to the steering of ships in America in about 1850. Hydraulic steering systems based on feedback were made for the British Royal Navy in about 1870. A steering system using an electrical servomechanism is represented in figure 18.8. A servomechanism

Figure 18.8 If a ship is not heading in the direction set on its gyrocompass, the error is determined by the potential divider, and the servoamplifier and motor correct the position of the rudder.

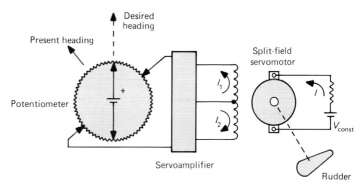

is a feedback system which uses a sensing element, an amplifier, and a servomotor to convert a low-powered mechanical motion into a mechanical motion requiring considerably greater power. In the system shown, the resistance coil of a potential divider is fastened to the ship's frame. The desired heading is determined by the gyrocompass setting. Any difference between the desired heading and the heading at any particular moment is signaled, in sign and magnitude, by the voltage the potential divider supplies to the amplifier. The rudder is positioned by a split-field servomotor (an electric motor that supplies power to a servomechanism). This is controled by varying the field current and maintaining a constant armature current. The motor's field coil has a center tap and the current in the two sections is supplied by the push–pull servoamplifier. Motion of the output is produced by an unbalanced current through the field windings.

(v) Voltage Control   A simple voltage control system for a DC generator is shown in figure 18.9(*a*). By adjusting the setting of the potential divider the output voltage of the generator is changed. Obviously this is an open-loop control: it does not respond to variation in load and for the usual compound-wound generator the output voltage will decrease somewhat with increasing output current.

We could 'close' the loop by having a human operator watch a voltmeter and continually adjust the potential divider. A more humane improvement is to return a portion of the output and compare it with the reference voltage, as in figure 18.9(*b*). Any difference signals a correction. The no-load voltage is adjusted by using the rheostat R in series with the shunt or bias field. If the terminal voltage begins to drop, the difference in voltage between the reference and output forces a current through the second or control field, causing the generator to maintain the desired voltage.

A simple electronic voltage regulator using feedback is shown in figure 18.10. It is designed to hold the voltage across the load resistance R_L constant within a small tolerance. Suppose either

(*a*)

(*b*)

Figure 18.9 The desired generator voltage is obtained in the open-loop system (*a*) by setting the potential divider to supply the appropriate field current. In the closed-loop system (*b*) variation in load voltage produces a correction through change of current in the control-field coil.

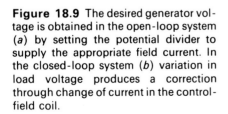

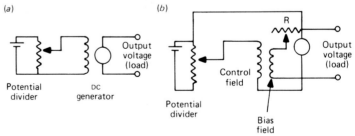

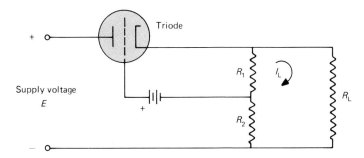

Figure 18.10 Voltage across the load resistance R_L is held constant through change in bias voltage of the triode.

the supply voltage E or load resistance R_L decreases. Then the voltage across R_1 drops, the grid becomes more positive and the load current I_L increases. This tends to bring the load voltage back to the desired value.

(vi) Amplifier Feedback To introduce feedback into an electronic amplifier a fraction of the output signal is connected to the input. An amplifier will usually have better frequency response when the system has feedback. To analyze an amplifier with feedback consider figure 18.11. The magnitude of the sinusoidal input signal is E_s and the amplifier gain is A, which is a function of frequency. If there is no feedback, $E_o = AE_s$. If there is feedback, the input is $E_i = E_o + E_f$. The feedback term is $E_f = BE_o$, where B is the fraction of the output signal returned. So the overall gain of the system is

$$\text{Gain} = \frac{E_o}{E_i} = \frac{A}{1 - AB}. \tag{18.8}$$

The term $(1 - AB)$ is a complex number, so the magnitude and phase angle of the gain of the amplifier with feedback will differ from the gain without feedback.

At the frequency for which $AB = 1$, the denominator in equation (18.8) will be zero and the numerator not zero. If this occurs, the amplifier will oscillate at approximately the frequency for which $AB = 1$. Further, if there is a frequency for which $|AB| > 1$ and the phase angle of AB is zero or an integral multiple of $360°$, the amplifier will oscillate.

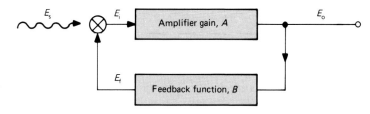

Figure 18.11 Feedback generally improves operation of an amplifier but may also lead to its oscillation.

(vii) Biological Feedback Control There are many types of biological control which use feedback; some are classed as voluntary, some as involuntary. The ears serve as sound sensors when you wish to orient your head to a sound—a voluntary control. In the finger-pointing loop, voluntary contraction of muscles depends on visual information.

An example of involuntary control is the adjustment of the pupil of the eye in response to different levels of illumination. In another example, chemoreceptors (chemical sensors) continually monitor the sugar concentration in the human body. If the concentration rises above the normal range (about 70–100 mg/100 ml) they signal the pancreas to deliver insulin to the bloodstream to process the sugar and thus lower its concentration. When this control loop fails, the condition called diabetes results.

These examples show that feedback control is a concept applicable in all fields of physics and engineering. You could study this subject on a get-acquainted level in a one-semester course, or go deeply into it and make a lifelong career of control theory and its applications, as many people have done. As our final model of an across-the-board idea, control theory stands unsurpassed.

▶ **Exercises**

18.1   Water is pumped out of the ground by a reciprocating piston pump and fed into the input end of a long hose having elastic walls. Describe the flow at different distances along the hose by a series of qualitative graphs. Assume that the flow at the input is the rectified half of a sine wave.

18.2   Relate the conditions and results of Exercise 18.1 to the flow of blood around the circulatory system where the walls of the aorta and major arteries are elastic.

18.3   In equation (18.7), suppose the minority particles did not recombine but had infinite lifetimes. How would the equation so modified compare with equations (18.6) and (18.4)? Identify the corresponding quantities.

18.4   A person carrying a nearly full cup of coffee may spill less if he does not watch the cup. Discuss this observation with respect to feedback control.

18.5   Draw a block diagram showing the kind of sensing element and control agent you would use (a) to adjust and control the power level of a nuclear reactor, (b) to maintain constant the heating value in Btu/ft^3 of gas supplied for home heating when that gas is made by mixing natural gas and butane, and (c) to control the temperature in an electrical home refrigerator.

18.6   For each of the following systems, draw functional block diagrams identifying the input energy or material, the process, the disturbances, the measuring device, the controler, and the final control element: (a) an automobile-engine cooling system; (b) body-temperature regulation in a warm-blooded animal; (c) a person driving a car which is coasting down hill on a winding road; (d) a continuously-rolling steel mill; (e) traffic control on a highway with access ramps.

18.7   A man buries his water-supply pipe 1 m deep, hoping that it will not freeze in winter. His hope is fulfilled when 21 December comes and goes, with the water still running. He is greatly upset and puzzled when the pipe freezes twenty days later. Explain this happening and calculate how much longer he will have to wait before the pipe thaws out. Assume the surface temperature of the ground follows a sinusoidal annual fluctuation with a mean temperature of $5\,^\circ C$ and an amplitude of $20\,^\circ C$.

18.8   Assume a sinusoidal flow of moisture into and out of the surface of porous ground, as suggested in figure 18.12. Set up and solve the flow equation. Compare it with equation (18.6) and identify the corresponding quantities.

18.9   The gain of a control system is defined as the ratio between output and input in the steady state. (a) Show that for the Watt governor (figure 18.3) the height of the revolving weights depends on speed only. (b) Treating the change in height as the change in output, calculate the gain in a Watt governor when the speed changes from 50 to 75 rev/min; and again when the speed changes from 250 to 275 rev/min. Comment.

18.10   At depths of 2, 4, and 8 m the annual ranges of temperature fluctuations are 5.6, 2.8 and $0.7\,^\circ C$. Show that the velocity of propagation of temperature waves into the Earth is 18.1 m yr^{-1}.

18.11   In Exercise 18.10, what is the amplitude of the temperature fluctuation at the surface? What are the phase lags at the three depths given?

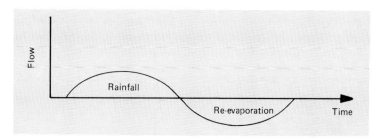

Figure 18.12

Answers to Exercises

Chapter 2   2.7   30 tons in the year 2000
                2.14   (a) The day before, (b) four days later

Chapter 3   3.1   8700 yr
                3.2   7.22 PM
                3.6   41% of its starting value
                3.8   35%

Chapter 4   4.5   3.9%
                4.6   1.0 mole l^{-1}
                4.8   1.4 cm^{-1}
                4.9 1.5×10^{14}
                4.13 $-19.9°$ and 19.998 °C
                4.14   5 W

Chapter 5   5.3 $h_{\text{final}} = I/f\rho g$
                5.6 $\tau = 0.001$ s
                5.8   (a) 8 m s^{-1} (b) 20 s
                5.9   Taking the low value for R the US would need to decrease
                        consumption by 15% per century

Chapter 6   6.6   (a) Four times, (b) twice, (c) no change
                6.11 $[\sqrt{(\rho_e g/\rho L)}]/2\pi$
                6.12 $(1/CR^2)^{1/2}$
                6.14   (a) $a = 10\,e^{-0.04606\,t}$ (b) 15.61 s
                6.18   (a) b/a (b) $-a^4/b^3$ (c) $2\pi\sqrt{[a^4(m_1 + m_2)/b^3 m_1 m_2]}$

Chapter 7   7.10   (a) 32 cm (b) 8.0 cm (c) 41.7 m s^{-1} (d) 1.30 Hz (e) 0.768 s
                7.12   (a) 0.20 cm (b) 5.0 cm (c) 60 cm s^{-1} (d) 12 Hz (e) -0.018 cm
                7.13   1.35 W m^{-2}

Chapter 8   8.1   (a) $y_b = 2a \cos(2\pi x_b/\lambda) \sin(2\pi vt/\lambda)$
(b) $v_y = dy/dt = 4\pi va/\lambda \cos(2\pi x_b/\lambda) \cos(2\pi vt/\lambda)$

8.2   458, 917, 1375, 1833, ... Hz.

8.4   3872 N

8.5 $T_n = 0.6 \times 10^{-8}/n\,\text{s}$

Chapter 9   9.1 $6.6\,\text{m s}^{-1}$

9.4   7

9.5   1.00074

9.6 $7.25 \times 10^{-5}\,\text{cm}$, $4.83 \times 10^{-5}\,\text{cm}$

9.7 $0.41\,\text{s}^{-1}$

Chapter 10   10.6   About $0.7°$

10.8   1.1 mm

Chapter 13   13.7 $u = d\omega/dk$

13.8   (a) $u = v - \lambda(dv/d\lambda)$—Rayleigh's formula   (b) $u = v\left(1 + \dfrac{\lambda dn}{nd\lambda}\right)$

13.9   (a) $u = v$ (b) $u = v/2$ (c) $u = 3v/2$ (d) $u = 2v$ (e) $u = c^2/v$

13.10 $u = 1.115c$

Chapter 14   14.1 $8.6 \times 10^3\,\text{m s}^{-1}$

14.2 $11.6 \times 10^3\,\text{K}$

14.3   (a) 15 keV; (b) 6.2×10^{10} atm

14.4   13.5 ft

14.5 $3.4 \times 10^6\,\text{s}^{-1}$

14.7   (a) 5536 m (b) 5536

14.9   0.36

Chapter 15   15.4 $3.37 \times 10^{-8}\,\text{s}$, 2.10×10^{10} electrons

15.7   about 5800 K

15.8   Yes, barely

15.9   About 3 days

Chapter 16   16.8   About 20 Å

16.9 $0.014\,\text{m s}^{-1}$

Chapter 17   17.9   (a) $[ML^2]$ (b) $[FLT^2]$ (c) $[L] = [T]$, $[\text{velocity}] = 1$

17.10   (a) $T = CRf(CR^2/L)$

17.11 $a = \omega^2 r$

17.12 $v \propto \sqrt{(l/g)}$

Chapter 18   18.9   (b) 31 %, 4 %

18.11 $11.2°$; 0.93, 1.38, 2.77 rad

Further Reading

Chapter 1   **Bassett C R** and **Pritchard M D W** 1967 Educational analogues for the study of intermittent heating *Physics Education* **2** 44–6
Cheng D K 1959 *Analysis of Linear Systems* (Reading, MA: Addison-Wesley) chs 4, 8
Cowan J D and **Kirschbaum H S** 1961 *Introduction to Circuit Analysis* (Columbus, OH: Charles E Merrill) ch. XII
Duhem P 1954 *The Aim and Structure of Physical Theory* (Princeton: Princeton University Press) ch. IV
Eckert E R G and **Goldstein R J** (ed.) 1967 *Measurements in Heat Transfer* 2nd edn (New York: Hemisphere and McGraw-Hill) pp 397–423
Harman W W and **Lytle D W** 1962 *Electrical and Mechanical Networks* (New York: McGraw-Hill)
Leatherdale W H 1974 *The Role of Analogy, Model and Metaphor in Science* (New York: American Elsevier) Extensive bibliography
Murphy G 1950 *Similitude in Engineering* (New York: Ronald Press)
Murphy G, Shippy D J and **Luo H L** 1963 *Engineering Analogies* (Ames, Iowa: Iowa State University Press) §100–§700
Polya G 1957 *How to Solve It* 2nd edn (New York: Doubleday) Note the chart, pp xvi, xvii on how to solve it
——1962–5 *Mathematical Discoveries; on Understanding, Learning and Teaching Problem Solving* (New York: Wiley)
Segrè E 1972 Major problems in contemporary physics *Endeavour* **31** 55–5

Chapter 2   **Bartlett A A** 1976 The exponential function   parts I and II *The Physics Teacher* **14** 393–401, 518–9
——1977 The exponential function—part III *The Physics Teacher* **15** 37–8
——1978 The exponential function—parts IV and V *The Physics Teacher* **15** 98, 225–6
Killingbeck J and **Cole G H A** 1971 *Mathematical Techniques and Physical Applications* (New York: Academic Press) §4-8

Malthus T R 1960 *On Population* (New York: Modern Library)

Meadows D H *et al* 1972 *The Limits of Growth* (Potomac Associates)

Youse B K and **Stalnaker A W** 1969 *Calculus for the Social and Natural Sciences* (Scranton: International Textbook) §97

Chapter 3   **Bartlett A A** 1976 The exponential function—part I *The Physics Teacher* **14** 393–401

Bateman H 1910 *Proc. Cambridge Phil. Soc.* **15** 423

Johnson R E and **Kiokemeister F L** 1964 *Calculus with Analytical Geometry* 3rd edn (Boston, MA: Allyn and Bacon) 273–4

McGraw-Hill 1977 *Encyclopaedia of Science and Technology* vol. 4 (New York: McGraw-Hill) 143–5

Rutherford E 1913 *Radioactive Substances and their Radiations* (Cambridge: Cambridge University Press). This classic book treats the exponential decay of radioactive substances in successive transformations. The general theory is given also by Bateman (1910).

Sutton O G 1962 *Mathematics in Action* (London: G Bell and Sons) pp 49, 223

Chapter 4   **Blanchard C H** *et al* 1969 *Introduction to Modern Physics* 2nd edn (Englewood Cliffs, NJ: Prentice-Hall) p 349

Cork J M 1933 *Heat* (New York: John Wiley) p 106

Weber R L 1950 *Heat and Temperature Measurement* (Englewood Cliffs, NJ: Prentice-Hall) p 40

Weidner R T and **Sells R L** 1973 *Elementary Modern Physics* (Boston, MA: Allyn and Bacon) pp 132, 278

Worthing A G and **Halliday D** 1948 *Heat* (New York: Wiley) ch. 7

Chapter 5   **Daly H E** (ed.) 1977 *Toward a Steady-state Economy* (San Francisco: W H Freeman) chs 1, 3

Chapter 6   **Barker J R** 1964 *Mechanical and Electrical Vibrations* (New York: Wiley) An analytical discussion of oscillating systems from an engineering viewpoint.

Bishop R E D 1965 *Vibration* (New York: Cambridge University Press) A general account of vibrations and engineering problems based on the Christmas Lectures at the Royal Institution, London.

Davies B 1978 Mathematical models in oscillation theory *Physics Education* **13** 282–6

Firestone F A 1956 Twixt earth and sky with rod and tube; the mobility and classical impedance analogies *J. Acoust. Soc. Am.* **28** (6) 1117–53

——1957 The mobility and classical impedance analogies *American Institute of Physics Handbook* (New York: McGraw-Hill) **3** 140–79

French A P 1971 *Vibrations and Waves* (MIT Introductory Physics Series) (New York: W W Norton)

Jones B K 1973 Parametric oscillators and amplifiers—parts I and II *Physics Education* **8** 310–14, 374–6

Morrill B 1957 *Mechanical Vibrations* (New York: Ronald Press) ch. 8

Olson H F 1958 *Dynamical Analogies* (New York: Van Nostrand) §V, §VI, §XI, §XIII, §XIV. An excellent text presenting analogies among electrical, mechanical rectilineal, mechanical rotational, and acoustic systems.

Rocard Y 1960 *General Dynamics of Vibrations* 2nd edn (New York: Frederick Ungar) ch. 2 §19

Sharman R V 1966 Vibration, wave motion, and sound *Physics Education* **1** 271–5

Skoglund V J 1967 *Similitude: Theory and Applications* (Scranton: International Textbook)

Thompson W T 1964 *Mechanical Vibrations* (London: Allen and Unwin)

Waldron R A 1964 *Waves and Oscillations* (New York: Van Nostrand)

Chapter 7

Benade A H 1976 *Fundamentals of Musical Acoustics* (New York: Oxford University Press)

Beranek L L 1954 *Acoustics* (New York: McGraw-Hill)

Bondi H 1961 Gravitational waves *Endeavour* **20** 121–30

Brillouin L 1953 *Wave Propagation in Periodic Structures* (New York: Dover)

Ghatak A K 1972 *An Introduction to Modern Optics* (New York: McGraw-Hill)

Jenkins F A and **White H E** 1976 *Fundamentals of Optics* 4th edn (New York: McGraw-Hill)

Kay R H 1976 The hearing of complicated sounds *Endeavour* **35** 104–9

Kock W 1965 *Sound Waves and Light Waves* (Garden City, NY: Anchor Books)

Mason E A *et al* 1971 Rainbows and glories in molecular scattering *Endeavour* **30** 91–6

Meyer-Arendt J R 1972 *Introduction to Classical and Modern Optics* (Englewood Heights, NJ: Prentice-Hall)

Pierce J R 1974 *Almost All About Waves* (Cambridge, MA: MIT Press)

Reid J S 1976 Phonon gas *Physics Education* **11** 348–53

Shive J N *Similarities in Wave Behavior* (Garden State: Bell Laboratories and Novo)

Wood R W 1934 *Physical Optics* 3rd edn (New York: Macmillan)

Chapter 8

Brown A F 1976 Seeing with sound *Endeavour* **35** 123–8

Morse P M *Vibration and Sound* (New York: McGraw-Hill)

Pierce J R 1974 *Almost All About Waves* (Cambridge, MA: MIT Press)

Rushton W A H 1967 Effect of humming on vision *Nature* **216** 1173

Sharman R V 1966 Vibration, wave motion and sound *Physics Education* **1** 271–5

Sutton O G 1962 *Mathematics in Action* (London: G Bell & Sons) ch. 4 pp 91–130. This volume contains a simple explanation of the part played by mathematics in selected topics in applied science.

Towne D H 1967 *Wave Phenomena* (Reading, MA: Addison-Wesley)

Williams P C and **Williams T P** 1972 Effect of humming on watching television *Nature* **239** 407

Chapter 9　**De Witte A J** 1967 Interference in scattered light *Am. J. Phys.* **35** 301–13

Durelli A J and **Parks V J** 1970 *Moiré Analysis of Strain* (Englewood Cliffs, NJ: Prentice-Hall)

Feynman R P, **Leighton R B** and **Sands M** 1964 *The Feynman Lectures on Physics* vol I (Reading, MA: Addison-Wesley)

Françon M 1966 *Optical Interferometry* (New York: Academic Press)

Gabor D and **Stroke G W** 1969 Holography and its applications *Endeavour* **28** 40–7

Givens M P 1967 Introduction to holography *Am. J. Phys.* **35** 1056–64

Jaffe B 1960 *Michelson and the Speed of Light* (New York: Doubleday-Anchor)

Jenkins F A and **White H E** 1957 *Fundamentals of Optics* 3rd edn (New York: McGraw-Hill)

Martienssen W and **Spiller E** 1964 Coherence and fluctuations in light beams *Am. J. Phys.* **32** 919–26

Meyer-Arendt J 1972 *Introduction to Classical and Modern Optics* (Englewood Cliffs, NJ: Prentice-Hall) §2.5, §2.6, §2.8

Michelson A A 1903 *Light Waves and Their Uses* (Chicago: University of Chicago Press)

Oster G 1968 Moiré patterns in physics *Endeavour* **27** 60–4 Survey of applications, reference to kits available from Edmund Scientific Co., Barrington, NJ

Chapter 10　**Beranek L L** 1954 *Acoustics* (New York: McGraw-Hill) ch. 4

Cosslett V E 1966 *Modern Microscopy* (London: G Bell and Sons) ch. 4

Erikson K R, **Fry F J** and **Jones J P** 1974 Ultrasound in medicine *IEEE Trans. Sonics Ultrasonics* **SU21** (3) 144–70

Farago P S 1974 Polarised electrons *Endeavour* **33** 143–8

Frank N H 1950 *Introduction to Electricity and Optics* 2nd edn (New York: McGraw-Hill) ch. 16

Holt C A 1963 *Introduction to Electromagnetic Fields and Waves* (New York: Wiley) ch. 23

Houston R A 1921 *A Treatise on Light* (London: Longmans) ch. 2

Jenkins F A and **White H E** 1976 *Fundamentals of Optics* 4th edn (New York: McGraw-Hill) ch. 1 §10

Kinsler L E and **Frey A R** 1962 *Fundamentals of Acoustics* 2nd edn (New York: Wiley) ch. 7

Kock W E 1965 *Sound Waves and Light Waves* (New York: Doubleday-Anchor)

Kraus J D 1973 *Electromagnetics* 2nd edn (New York: McGraw-Hill) ch. 13

Lorrain P and **Corson D R** 1970 *Electromagnetic Fields and Waves* 2nd edn (San Francisco, CA: W H Freeman) ch. 13

Meyer-Arendt J R 1972 *Introduction to Classical and Modern Optics* (Englewood Cliffs, NJ: Prentice-Hall) §1.2–§1.9

Ramo S and **Whinnery J R** 1944 *Fields and Waves in Modern Radio* (New York: Wiley) ch. 12

Stratton J A 1941 *Electromagnetic Theory* (New York: McGraw-Hill) pp 438–54

Swift J A 1970 *Electron Microscopes* (New York: Barnes and Noble)

Chapter 11   **Albert A L** 1943 *Fundamentals of Telephony* (New York: McGraw-Hill) ch. 10

Beranek L 1954 *Acoustics* (New York: McGraw-Hill) chs 26, 27

Brillouin L 1946 *Wave Propagation in Periodic Structures* (New York: McGraw-Hill)

Karakash J 1950 *Transmission Lines and Filter Networks* (New York: Macmillan)

Kinsler L E and **Frey A R** 1950 *Fundamentals of Acoustics* (New York: Wiley) ch. 8

Mott N F and **Gurney R W** 1964 *Electronic Processes in Ionic Crystals* 2nd edn (New York: Dover)

Page L and **Adams N I** 1931 *Principles of Electricity* (New York: Van Nostrand) ch. 15

Chapter 12   **Blume L F** *et al* 1951 *Transformer Engineering* 2nd edn (New York: Wiley) ch. IV

Boylestad R L 1968 *Introductory Circuit Analysis* (Columbus, OH: Charles E Merrill) ch. 22

Department of Electrical Engineering, MIT 1943 *Magnetic Circuits and Transformers* (New York: Wiley)

Firestone F A 1956' Twixt earth and sky with rod and tube; the mobility and classical impedance analogies *J. Acoust. Soc. Am.* **28** (6) 1117–53 [Article appears also in *American Physical Society Handbook* 1957 (New York: McGraw-Hill) pp 3–140]

Kimbark E E 1956 *Electrical Transmission of Power and Signals* (New York: Wiley)

Laithwaite E R 1969 Electromagnetic puzzles *Physics Education* **4** 96–100, 114–5 These problems are examples of the type which we tend to solve by simple analogies.

MacInnes I 1972 The lever as an impedance matching device *Physics Education* **7** 509–10

Millman J and **Halkias C C** 1972 *Integrated Electronics: Analog and Digital Circuits and Systems* (New York: McGraw-Hill) pp 684–5

Nordenberg H M 1964 *Electronic Transformers* (New York: Reinhold)

Skilling H H 1965 *Electrical Engineering Circuits* 2nd edn (New York: Wiley)

Chapter 13   **Bishop A R** and **Schneider T** 1978 *Solitons and Condensed Matter Physics* (New York: Springer Verlag)

Born M and **Wolf E** 1970 *Principles of Optics* 4th edn (Oxford: Pergamon)

Davidson C W 1978 *Transmission Lines for Communications* (New York: Halsted and Wiley)

Feynman R P and **Leighton R B** 1963 *The Feynman Lectures in Physics* vol. 1 (London: Addison-Wesley) ch. 31

Good R H Jr and **Nelson T J** 1971 *Classical Theory of Electric and Magnetic Fields* (New York: Academic Press)

Houston R A 1934 *A Treatise on Light* (London: Longmans-Green) ch. 24

Jenkins F H and White H E 1957 *Fundamentals of Optics* 3rd edn (New York: McGraw-Hill)

King R W P 1955 *Transmission-line Theory* (New York: McGraw-Hill)

Weisskopf V F 1968 How light interacts with matter *Sci. Am.* **219** (3) 60

Wood R W 1934 *Physical Optics* 3rd edn (New York: Macmillan)

Chapter 14
Dougal R C 1976 The presentation of the Planck radiation formula *Physics Education* **11** (6) 438–43

Feynman R P 1964 *Lectures on Physics* vol. 1 chs 39–44 (Reading, MA: Addison-Wesley)

Leighton R B 1959 *Principles of Modern Physics* (New York: McGraw-Hill) ch. 10

Loeb L B 1927 *Kinetic Theory of Gases* (New York: McGraw-Hill)

Sproull R L 1959 *Modern Physics* (New York: John Wiley)

Chapter 15
Barnes R B and Silverman S 1934 Brownian motion as a natural limit to all measuring processes *Rev. Mod. Phys.* **6** 162–92

Frank N H 1950 *Introduction to Electricity and Optics* 2nd edn (New York: McGraw-Hill) ch. 20

Furth R 1950 Limits of Measurement *Sci. Am.* **183** (1) 48–50

Jenkins F A and White H E 1976 *Fundamentals of Optics* 4th edn (New York: McGraw Hill) ch. 21

Kerker M 1974 Brownian movement and molecular reality prior to 1900 *J. Chem. Education* **51** 764–8

Kittel C 1958 *Elementary Statistical Physics* (New York: Wiley) ch. 19

Nelson E 1967 *Dynamical Theories of Brownian Motion* (Princeton, NJ: Princeton University Press)

Perrin M J 1910 *Brownian Movement and Molecular Reality* transl. F Soddy (London: Taylor & Francis)

Chapter 16
Ashkin A and Dziedzic J M 1973 Radiation pressure on a free liquid surface *Phys. Rev. Lett.* **30** 139–42

——1975 Optical levitation of liquid drops by radiation pressure *Science* **187** 1073–5

Frank N 1950 *Introduction to Electricity and Optics* 2nd edn (New York: McGraw Hill) §20-2

Hull G F 1905 Elimination of gas action in experiments on light pressure *Phys. Rev.* **20** 188, 292–9

Nichols E F and Hull G F 1901 A preliminary communication on the pressure of heat and light radiation *Phys. Rev.* **13** 307–20

——1903 The pressure due to radiation *Phys. Rev.* **17** 26–50, 91–104

Richtmyer F K, Kennard H and Lauritsen T 1955 *Introduction to Modern Physics* (New York: McGraw-Hill) p 108ff

Saha M N and Srivastava B N 1935 *A Treatise on Heat* (Allahabad: The Indian Press) pp 564–73

Smith R A 1970 Light scattering *Endeavour* **29** 71–6

Chapter 17   **Bassett C R** and **Pritchard M D W** 1968 The teaching of structural problems by the use of electrical analogues *Physics Education* **3** 242–5

Buchdahl H A 1970 *An Introduction to Hamiltonian Optics* (Cambridge: Cambridge University Press) A graduate-level text, with problems and solutions. The Hamiltonian method gives results governing behavior of lens systems with respect to *symmetries* they possess.

Bridgman P W 1931 *Dimensional Analysis* (New Haven: Yale University Press)

Gruber B 1980 *Symmetry in Science* (New York: Plenum)

Hesse M 1952 Operational definition and analogy in physical theories *Br. J. Phil. Sci.* **2** 281–94

Isenberg C 1975 Problem solving with soap films—parts I and II *Physics Education* **10** 452–6, 500–503 An analog system consisting of plates, pins, and soap films provides a simple method of solving difficult mathematical problems in two dimensions. Some practical applications discussed.

MacLeod A M 1975 An elementary introduction to TTL (transistor-transistor logic) *Physics Education* **10** 440–5

——1976 a second step in TTL circuitry *Physics Education* **11** 111–16

——1976 Arithmetic operations using TTL *Physics Education* **11** 336–40

Papoulis A 1968 *Systems and Transforms with Applications in Optics* (New York: McGraw-Hill)

Polya G 1954 *Mathematics and Plausible Reasoning*: vol. I *Induction and Analogy in Mathematics*; vol. II *Patterns of Plausible Inference* (Princeton: Princeton University Press)

Shubnikov A V and **Kopstik V A** 1974 *Symmetry and Science in Art* (New York: Plenum)

Streater R F and **Wightman A S** 1964 *PCT, Spin and Statistics and All That* (New York: W A Benjamin)

Sutton O G 1960 *Mathematics in Action* (New York: Harper Torch Books)

Wigner E P 1965 Violations of symmetry in physics *Sci. Am.* **313** (6) 28–36

Chapter 18   **Adler I** 1961 *Thinking Machines* (New York: John Day)

Barbe E C 1963 *Linear Control Systems* (Scranton: International Textbook)

Bruns R A and **Saunders R M** 1955 *Analysis of Feedback and Control Systems* (New York: McGraw-Hill)

Del Toro V and **Parker S R** 1960 *Principles of Control Systems Engineering* (New York: McGraw-Hill)

Doebelin E O 1962 *Dynamic Analysis and Feedback Control* (New York: McGraw-Hill)

Dorf R C *Modern Control Systems* (Reading, MA: Addison-Wesley)

Hadley W A and **Longobardo G** 1963 *Automatic Process Control* (Reading, MA: Addison-Wesley)

Hardie A M 1964 *The Elements of Feedback and Control* (London: Oxford University Press)

Martin H R 1968 *Introduction to Feedback Systems* (New York: McGraw-Hill)

Weber R L 1950 *Heat and Temperature Measurement* (Englewood Cliffs, NJ: Prentice-Hall) ch. 9

Wiener N 1948 *Cybernetics* (New York: Wiley)

Suggested Guides
to the Library

When you need to find information on a particular topic in physics—perhaps an introduction to a subject unfamiliar to you, or perhaps some numerical data—you may find it helpful to consult bibliographical guides such as the following:

Coblans H (ed.) 1975 *Use of Physics Literature* (London and Boston: Butterworths)

Guide to Reference Books 1st Edn 1902 (Chicago: American Library Association) A guide to reference materials in all disciplines. New editions and supplements are issued on a regular basis.

Parker C C and **Turley R V** 1975 *Information Sources in Science and Technology* (London and Boston: Butterworths)

Whitford R H 1968 *Physics Literature; A Reference Manual* 2nd edn (Metuchen, NJ: Scarecrow Press)

Walford A J (ed.) 1973 *Guide to Reference Material* 3rd edn (London: Library Association) This guide to all subject disciplines has one volume devoted to the sciences.

Index